Divine Science Bible Text Book by A. B. Fay

Library Home - A. B. Fay's Bio - More Authors - More Texts

THE DIVINE SCIENCE

BIBLE TEXT BOOK

Interpretation Based
Upon the Omnipresence of God

A. B. FAY, D. S. D.

"And ye shall know the truth and the truth shall make you free"

COLORADO COLLEGE OF DIVINE SCIENCE
DENVER, COLORADO

1

DENVER:
The W. F. Robinson Printing Co. 1920

2

THIS BOOK
IS DEDICATED TO MY FRIEND

FANNIE B. JAMES

WITH THE LOVE THAT LIVES ON IN THE
BEYOND, AS SURE AND ETERNAL AS
THE GOD WHO INSPIRES IT

3

Click on 🔄 **to return to the top of the Contents**

Need a daily spiritual boost?

Try DivineJournal.com

Another free resource to support your spiritual journey!

Preface

The literal interpretation of the Bible has done much towards driving thoughtful minds into unbelief. Knowing as we do that the oriental language abounds in symbolism, it should be easy for us to understand that we cannot insist upon the literalness of this book which comes to us from the Orient. Many things in the Bible show in themselves the impossibility of being rendered literally, and if any part is admitted to be figurative so may other parts be.

Then we do well to remember, as one has said, that the Bible does not make Truth any more than books written on electricity make electricity. Man discovered Truth and then wrote what he knew about it. In the Bible is recorded simply what men have learned about Truth. Truth never destroys Truth, the fuller vision includes all the lesser that has preceded. Throughout the ancient writings we have messages from all stages of spiritual development, from primitive man to man in the highest plane of unfoldment. We find in these writings the lispings of babes and the utterances of the fully developed or the wise man. "The first man (infant or first view of man) is of the earth earthy; the second man (truer understanding of man) is the Lord from heaven."

5

As long as man invents systems of Truth they will prove to be limited and partial visions of the Whole Truth, which is known only as we understand the Omnipresence of God. This, Omnipresence of God, means literally the omnipresence of all that God is; Love, Wisdom, Knowledge, Understanding, Power, Life and Joy.

Our faith must be logical to be scientific, and it must be scientific to be intelligent and reasonable. Truth is changeless, only opinions of Truth change.

After all, the Bible was written by persons like ourselves, whose greatest claim was not so nitwit that they knew God as that they longed for Him, and so longing, spoke out of their best and deepest understanding.

Whatever our forefathers saw or received we may see, and surely after these years of opportunities, we should be able to receive, see or perceive a deeper, broader, clearer vision. One says, "All times are Bible times, sacred times; all events are Bible events; all individuals are Bible individuals; all said in the Bible is a constant process."

So while Truth, the Inspirer, must be infallible, the words that express it are but faulty media of human thought. The Idea is heaven horn, but the words are man-made.

6

These words are just like a scaffolding which supports the workmen while they build a tower; it does not support the tower, neither do words support Truth, they help man to express or build up what Truth sees; and just as the scaffolding rises higher as the tower progresses, so as Truth grows in man's thoughts, does he speak higher words concerning it. At last when the tower is finished, the scaffolding is removed, that the building may be seen unobstructed by the temporary framework.

So, too, when Truth is fully realized, the tangle of words is not needed, their work is done. The finishing touch has been put on Truth's temple, not by words, for these only served as stepping stones to its highest pinnacle, but by unfoldment into consciousness. The completed Truth has been reached by man's thought.

When we hear Bible criticism of today, we have no anxiety; it is just the work of tearing down the scaffolding upon which men have stood to reach the heights of Truth, and now the completed temple can do without the scaffolding. There is no danger that the building will fall when it is removed, for we know that Truth never rested upon words, never depended upon any human concept. "When that which is perfect is come, then that which was in part shall be done away."

7

Let us tear away the scaffolding, and look at the beauty of the perfect building in the new light of today. Let us put the thought of man's early spiritual experience into the "new tongue."

While the following lessons are written for all, all may not be ready for them. Consider this advice: Let a new student delay the study of this course until he has demonstrated the Principle of Divine Science as learned in the preceding College Courses. In fact, faithful study and work must have been done to prepare one for what is given here.

These Bible lessons are published now for the first time. They are given in the Colorado College of Divine Science as an advanced Bible Course for students who for years perhaps have demonstrated the Truth of the Omnipresence of God. They involve deep Truth which can he received only by those prepared for it. To those who have been students of Truth we say, Prepare yourselves for this advanced study. You will need to enter into it as you entered your first class with unbiased opinions and unprejudiced thoughts.

In the study of Divine Science all avenues of thought are considered willingly and lovingly. We believe that the person who denies the existence of anything beyond the horizon of his understanding, because he cannot make it harmonize with his accepted opinions, is as unstable as he

8

who believes everything without any discrimination.

While we believe in repetition, knowing how it deepens impressions, and we repeat often in this course, yet each lesson has its individual purpose, and the first chapters will elucidate much in the last; therefore we ask our students to study this text-book on the Bible consecutively. We ask them to study this book for it is the faithful interpretation of that great human document, the Bible, from the basis of the Omnipresence of God. It is the evolution of years of practice and study, its depth cannot be probed by a single reading.

It is our high hope that the simplicity of its presentation will be the means of tearing away forever the ecclesiastical bandages that have been bound about the man Jesus, hiding from the longing hearts of men the Truth of his loving message. He taught us to say "our Father," and we find in Jesus' life the mightiest influence to lift our aspirations to his Father and our Father, his God and our God, as well as the revelation of peace, hope and immortal life needed by us all for the perfect unfoldment of our being. Through this understanding of his mission the most barren life catches a glimpse of that great immortal life that is our's today.

It is only when we comprehend how deep, rich and full are the utterances of Jesus that we begin

9

to recognize from what a deep consciousness they must have come.

The marvellous beauty and power of Christianity can be truly realized only when Jesus' teachings are regarded as statements of Truth to be lived now; not as prophecies relating to some far-off spiritual existence.

"The kingdom of Goal is within you," is the truth he sought to impress upon men. We are designed to live the spiritual life now, and to reap its harvest of power and happiness, now.

In Jesus the perfect ideal has forever become the actual; the exhibition to the world of a perfect humanity in Jesus' life has taught one lesson which our systems of theology have overlooked. It has demonstrated that sin and disease are no part of man, they are not constituents of human nature, but that health and harmony are the sure resultant of Law understand and obeyed.

Jesus compassed the length, breadth, depth and height of Life. He saw Life steadily and saw it whole, and in his presence and at his word the water of Life, in all its vessels of love and labor, culture and religion, became wine. And today the spirit of his great love is drawing the world to God.

We sincerely hope that natural scientists, ever on the alert to hear the word, "Go forward," will find in these pages the help these lessons have

10

given to other students and will glorify the Divine Source of all revelation, rather than any expression of it.

It is necessary for the student to look up every marginal reference before reading the note upon it. When other than the King James version of the Bible is used the reference is given.

We have purposely avoided the use of names throughout this volume, in order that the attention of the student might not be withdrawn from the subject. Quotation marks alone will indicate thoughts taken from others.

We are indebted for many of the historical references, to the Dictionary of the Bible, edited by James Hastings, M.A., D.D.

For the spiritual interpretation of the Bible herein expressed, the real aim of the work, we gratefully and lovingly acknowledge our deep indebtedness to our teachers, Mrs. M. E. Cramer, the founder of Divine Science, and Mrs. Fannie B. James, former President of the Colorado College of Divine Science. THE AUTHOR.

11

Return to the Top  Return to the Top

CHAPTER I. - GENESIS ---

A. "The study of the genesis of creation requires the very closest analysis and adherence to principle for its understanding and application. For all scientific analysis must proceed from principle, and all demonstration be obtained through its application.

"Since theology is the science of God and his relation to his creatures; the science which treats of the existence, character and attributes of God, His laws and Government; the doctrines we are to believe, and the doctrines we are to practice; the truth of this course in Bible Study is to be found in the study of God who is the Whole, not in the study of circumstances and events of the past and the committing to memory of their dates. There is one God, infinite, omniscient, omnipotent, and omnipresent, whose presence being everywhere secures equal power and knowledge to His creatures and proves that He is not a respecter of persons.

"In Genesis --- the true account of creation --God creates everything that is created, and there is nothing apparent but God and His word; nothing

15

---

more nor less, than God and Him manifesting Himself.

"No attempt is made to explain who and what God the speaker is, who precedes and is the Source and Cause of Creation.

"The Source and Cause is designated by the term Elohim, a plural noun which means inherencies, invisible powers, and is rendered God. It is understood that God is, in the beginning, and that all power necessary to create, and all idea of Creation is possible with God and is God."

WHERE AND WHAT IS THE BEGINNING?

"In the beginning --- cause and source --- was the Word, and the Word was with God, and the Word was God.

John 1: 1

John 17:5

"So the things that are made are not made out of nothing. The claim that something is made out of nothing is without foundation, hence unthinkable, and cannot be the Creator's method. "The method of Divine Mind, which is that like produces like, must be the only law by which wholeness can be maintained. That is the law of the Holy One; is the law by which God's infinitude is forever maintained and which Jesus declared to Nicodemus when he said, That having been begotten of Spirit is Spirit.

"God creates not in time but eternity; He creates not in place but within Himself. Thus it

16

---

is seen that we live and move and have our being in Him.

"The word 'genesis' means the act of producing, giving birth or origin to anything, the process of originating or forming. The book is the record of an eternal method, a never ending action. Its first statement explains the origin of all existence and proclaims God, heaven and earth, to be a trinity in unity, co-eternal as Creator, Creative action and Creation.

"As true as the time honored mathematical axiom that the whole is equal to the sum of all its parts, so is it true that God is equal to the whole --is totality.

"As in the science of numbers all figures are the repetition of one and that one repeated appears as different forms, giving a variety of figures each representing the number of times the one has been repeated, so in

Divine Science the starting point is the unit, the One God. The symbol of the Divine method of Creative action is to be found in the science of numbers; all forms, simple and complex, are the expression of the One --- they exist as the One revealed and are 'members one of another.' Hence the variety of living forms is as numerous as is the unlimited possibility of the Creative principle to produce them, which is the unlimited power of God to express Himself.

17

Matt 5 48

Col. 1 28

Eph. 4: 6

"Each living form is as perfect within itself as is the One producing it; each form stands for one Life Principle, for the one God and Father of all, who is above all, and through all, and in you all.

"Existence, then, is not here on its own account, but is here because God the Creator is here.

"In the truth of the omnipresence of the Creator is the foundation for all scientific explanation through the law of expression, for there could be no expression anywhere without the presence of an. Expressor. It is clear then that creating is the work of self expression, that the Being who expresses is before expression on takes place and is also the beginning of the same.

"Since God alone is in the beginning, it would not be true analysis from Principle to suppose that He made creation out of nothing; that the Infinite Mind makes creatures out of nothing is not true, but creates within Itself, since Mind thinks within and unto Itself.

"I am the beginning and the end, was spoken by the same Mind that speaks things into existence, and which knows that nothing can be caused to exist in Mind that is not there before it is formed. All creation is in the Infinite Mind before it is expressed, and can say, I am the beginning.

"It is self evident that the beginning is in God and is God's creative action. - Observation has not

18

---

revealed this truth to us, neither can it he comprehended by observation, but Divine perception perceives, and consciousness realizes it to be the Truth of Eternal Being.

"The new and living way --- Divine Science --shows us that we must start from Spirit, and not from the 'natural,' in our thinking.

"Since God is before He creates, there is no attitude of knowledge for us to take except that we are in the Universal Being before we are expressed in visible form; hence, I am before I think. It is evident that I, Divine Mind or God, create within Myself, unto Myself, and that creation lives in Me, and has being in Me and is Myself made manifest, and unto `Me,' all things live, invisible and visible in heaven and in earth. What is made is Mine."

Ex. 19:5
Deut. 10:14
Ps. 24: 1
Isa. 48: 13

The teaching which claims that visible things do not really evict, that they are not before us as we think, that we do not see them, but only think we do, that they really are only in our minds, does not understand that it is dealing lightly with a wonderful truth.

Verily all things are in Mind from eternity to eternity. No living thing is expressed that is not potential in Mind before it is expressed, and no thing is made with hands that is not in thought before it is made. Nothing is made or shaped from without, but everything is shaped or formed

2 Tim. 1: 9

19

---

within Mind; for this reason Mind sees visible objects, It sees and knows what is manifest within and unto Itself from the least to the greatest, from the sparrow to man.

Ps. 19: 1, 2

Rom. 1: 20

Acts 17: 28

Its omnipresence enables It to be the "all seeing I" --- the I of all --- hence Mind sees the whole universe which exists within It; and the whole is not merely intelligent, but is intelligence, not merely substantial and enduring, but is eternal substance. There is no account of any material creation, for the reason that there is no material Creator; God is Spirit.

Infinite Mind, Holy Spirit, Supreme Being, are used in. Divine Science as synonymous; there is no other Mind, no other Spirit, no Being besides Me is God's own claim. So the record of Creation is wholly spiritual and heavenly in its nature. The most stupendous truth reached as yet by natural science or accomplished by secular knowledge is, "Ether no less than Force, is a mode of the first Cause, is a form of the energy of God."

Heb. 11: 1

In Divine Science the method of reasoning is wholly deductive; the truth of the Universal Omnipresence is established in thought and from this basis conclusions about the individual are reached. Such reasoning affirms the manifestation of the Invisible in the visible. When the relationship between the two is fully perceived we

20

---

know that even our environment is Supreme Being.

We speak of coming into the Divine Presence when we pray as if we could ever withdraw from that Presence.

"Were there no medium about us save the atmosphere, this alone, with its force within our bodies, would lock us about with an inconceivable vigor. Let us remember that the atoms of atmosphere are not matter, but force, and force that both knows and feels. Yet this atmosphere is really a little thing compared with that other medium which holds up the worlds and suns and systems, harder than adamant, more unyielding than the granite hills. This is our environment. In this we walk and leap, in this we carry on our paltry concerns with the powers of God within us." The same truth that will enable us to analyze and understand Genesis will find readjustment and translation of all things; will find the truth of redemption to be the truth of Creation.

There is no conflict between the Mosaic account of creation and the gospel of St. John; or the conception and birth of Jesus, according to the gospels of Luke or Matthew. "In the beginning" refers not to time but to an eternal Creative action. It was with God and is God.

2 Cor. 5:17

Rom. 8:19-25

When is the beginning?

Genesis usually is described as implying a

21

---

work of the past, an act accomplished in a set time and once for all. Divine Science declares that God is yesterday, today and forever the same, hence what has been done by Him, is that which is being done now, and that which will be done throughout eternity. The creative week of Genesis denotes equally, therefore, the past, the present and the future. The record that sets forth the truth of God's work must be true for all time, which is the eternal present.

Rev. 3: 14

1 John 5:11-12

"Were the Bible account of the genesis of creation not the record of God's process of self expression today, it would he merely an account of a past event that belonged to time and place and not a living truth belonging to eternity, hence not a true record at this time. The faithful adhere to the principle of at-one-ment and know themselves to be with God in the beginning, that `I and my Father are one' now. The true know that truth is actualized within them from the beginning, from the first expression of eternal Being. They witness it here and now within. If the beginning were not now there would be no creation now and no beginning could be witnessed.

"No individual is conscious of having had a beginning because he is co-eternal with and has being in God; but every individual is conscious that the beginning of every accomplishment is in

22

Mind or Being, and that it is through Mind's action that Its idea is thought out and then wrought out or actualized. God creates to make --- or Mind thinks in order to form.

"Now that our perception has out-passed time and place, we perceive that the creative act is an eternal one, so the beginning is what the Infinite Mind is doing here and now   1 John 2:13-14
within Itself.

"No false supposition can enter here no opinions have anything to do with Creating; no reasoning by comparison can possibly take place in the light of pure intelligence, --- the day in which heaven and earth is created. Infinite space is peopled with God's idea of Himself, with an infinite variety of forms, all holding the same individual relationship to Him."

Creation can have Being nowhere but in that which is in the beginning. Being is Spirit,
Mind or God a thing cannot exist or be formed without a beginning; it cannot exist apart   Col.1.: 15-16
from its beginning in Source or Cause. `` The Creative Story is the alphabet of Science."
Science and religion spell out different messages to man, but they start with the same   Eph. 3: 14-15
alphabet. The first chapter of Genesis as a whole is a prophecy and its fulfillment; the
record carries us from Infinite Basis to Infinite Expression. The first chapter gives the outline of this wonderful process, and the rest of the Bible gives the details.

23

"As the artist begins his painting by outlining his mental picture on canvas, then proceeds to work in the detail of his ideal unto its completion, so in Genesis the first statement of the first day is one inclusive of the whole of creation, and what follows is descriptive of heaven and earth which are created in God in the beginning. They are simultaneous creations and include all the 'Hosts of them.'

"Days in Genesis have a spiritual meaning; the things of the Spirit refer to Spirit, they have no reference to periods of time, but rather to God's accomplishments. This is seen in that He called the firmament heaven and the dry land earth. We are to conclude that as all creation is in the omnipresent One, before it is expressed, its expression is simultaneous everywhere at the same time. The evidence we have of this is that both the visible and invisible exist at this time.

"All things are made by My word, and without it is not anything made that is made. The Creator is God, the Creative action is God and Creation is God. All existence is in God after it is expressed, as it is before. The earth never ceases to be without form and void before it is expressed, and is never apparent in anything but form when it is expressed."

The meaning of the word of God is form; the universe is the utterance of life, the spoken word.

24

The face of the deep is never the Source of light; there is never any light upon the face of the deep except that which is spoken forth --- or is let be manifest --- by the One who is light and in whom there is no darkness.

25a

Return to the Top  Return to the Top

B. God, Eternal Being, scientifically considered - p. 25

B. All creation is God expressed, as all figures are repetitions of One. Genesis is scientific truth, the science of God in Creation. There is no missing link in creation. Law works from within out, not from without in; there is no creation by reflection; no evolution from a lower to a higher plane; no unfoldment from "mud to monkey and from monkey to man." God expressing himself; God creating; God manifesting, creation is the gospel of the trinity. I am before I create; the whole God; God creating heaven and earth.

To understand creation one must lay aside personal opinions and accept the Infinite's idea of it. This idea is ever revealed as the word of God and is the new heaven and the new earth, which are to be perceived, entered into and enjoyed in God only; for in Him they exist.

The record of Genesis given so long ago in Hebrew will ever do its perfect work for those who accept it as the record of God expressing His possibilities; it is in every sense a revelation of God to every one who views it with the Christ consciousness of Being, ever recognizing that "I and my Father are one."

Rev. 2: 3

2 Cor. 6: 16

25

In the Genesis of Creation, separate personalities from God and individuals unlike each other in their real natures are not referred to. Supreme Being and its law working within and unto itself and the results following are alone referred to.

Every event of each day's accomplishment stands for God's self revealment and in man the nature of the whole is made known as being the image and likeness of God. We are taken to the Source, and creation is taking place.

Gen. 1: 1

John 14: 20

1 John 2: 24

"In the beginning God created the heaven and the earth." The word "beginning" means commencement; God in action or creating. Creation is eternally the new heaven and the new earth. My body is always fresh from the hands of the Creator.

1 John 5: 11

John 17: 3

The Infinite Creator, Source and Cause, is eternal. It begins Its work within Itself, and within Itself (in the beginning); It creates heaven and earth, the whole of creation, which is God idea expressed or revealed. God is one, inclusive of beginning and result. One, means God, whose beginning consists in the totality and harmonious union of Creator, Creating or Creation.

Acts 4: 24

First means before anything else in time, space or rank. First Day stands for the state of wholeness, for totality; one God inclusive of His action and result. The root meaning of one is to unite. It is the one-ness of God, heaven and earth. "As it is in the beginning so is it now, and ever more shall be, world without end." "Heaven and earth are conjoined now and are in God now and he comprehends the whole."

"And the earth was without form, and void; and darkness was upon the face of the deep; and the spirit of God moved upon the face of the waters."   Gen. 1:2

The earth potential in God is without form until it is formed by a beginning, and through   John 1:1
creative action. There is never a time when it is not without form before it is expressed.
All form is expressed by the Spirit of God's moving upon the face of the waters. Face means surface --- the exterior.

Without form and void means not spoken into expression; resting as idea in the Infinite Mind. The creating of anything implies Divine activity, and this action is universal.

"And darkness was upon the face of the deep." "Deep" stands for God invisible, and "face," meaning the exterior, is never the source and cause of light. All light comes from within; the "within" is the deep from which comes forth all expression.

"Moving upon the face of the waters" denotes action, and is the expressing of innate potentialities, the causing of form to appear.

"And God said, let there be light."

Gen. 1: 3

27

To God, who is pure intelligence, all that is spoken appears in form, and the coming forth of His word is form and is the light of the whole surface; it is the light of the world; "the light that lighteth every man."

1 John 1:1-5   "And there was light," means God's word is light. This does not make the face the
Source and Cause of light, but proves that expression is the result of light, now we
see face to face.
1 Cor. 13 2

2 Tim. 1:9-10   "Let there be" means, let be that which is. "I

am all in all," all light of pure intelligence. When the eternal is made visible it lights
the whole face

of the deep. It is only in light that things are visible. The first word spoken signifies that in the beginning all heaven and earth are made manifest in light.

Let all be as it is, pure light of intelligence. Without the word, which is with God and is God, is not anything made that was made.

The creative action, or causing of His possibilities to appear, is the saying "Let there be light," the revealing of the invisible idea in the light of knowledge and wisdom.

Gen. 1- 4   "And God saw the light, that it was good; and God divided the light from the
darkness." Commanded it to shine out of the darkness --- it was in the darkness.
God, who is light, is the one who sees the light and sees that it is good. Darkness
Ex. 20: 21   stands for God's presence unseen. To the Infinite

28

One there is no law save His own nature, and It declares that like produces like. That which is begotten of light is light. It is the all seeing eye single to truth that sees the whole body --- all creation --- full of light.

"And God called the light Day and the darkness He called Night, and the evening and the morning were the first day." Gen. 1: 5

"And God called the light day," means God was manifested in His word. " Called" means,
He had proceeded forth as the light manifest, which is Himself manifest, is day --- the truth that all in all is God and God manifest is eternal day. There is nothing hid that shall not be revealed in the light of this day. All is one illumination; One Intelligence filling immensity.

This is the day of the Lord, the first, and means totality, in which heaven and earth are ever being created; the day that the Lord God makes the earth and the heavens, and every plant of the field before it is in the earth, and every herb of the field before it grows.

The light which He called day, shines in darkness to those seeking knowledge and wisdom from without --- observation --- who believe they are to be obtained from something outside of conscious Being, but the darkness of this method comprehends not the light. John 1: 10

29

Prov. 8:22-29

John 3:16-20

"I am the light of the world," means I am now this truth and must necessarily find it --- what I am, conscious Being. The term "day" in the Brahmanical narration, is spoken of as a period of time when Brahm creates, the activity by which He manifests Himself and night as inactivity. Inactivity is the non-recognition of the truth that God is manifesting here and now; the absence of the light of His manifestation.

"Evening" means to blend together, to mix, to mingle. "Morning" means coming forth, birth.

"Since God alone is in the beginning, it is God who is manifested from the beginning, and the mingling together and perfect unity of what God is and what God does from the beginning is the evening of the first day; and the morning which follows is the birth or appearance of a heavenly result."

As we pass from the first day to the second, we have in the meaning of the word "evening," the blending together of what God is, with what God does and in the meaning of "morning," we have what God calls and proceeds forth in --which is Heaven the second day.

Gen. 1: 6

"And God said let there be a firmament in the midst of the waters, and let it divide the waters from the waters."

The first day stands for what God is as Creator,

30

Creative action and Creation. The following days describe in detail Creative action and Creation, or what heaven and earth mean to God.

Since God is One and One is all of Being, two cannot stand for another being but does stand for the action of Being, hence, the second day describes God creating in the beginning from within Himself. The root meaning of two or second is to do," to repeat it is the process of continuous creation.

The "Firmament" is the fixed foundation; well established basis, a wide extent. The original does not convey the sense of solidity, but of the stretching extension of the infinite and omnipresent activity of God, which produces living things. There is nothing more firmly fixed than eternal creative action, the basis and beginning of which is an eternal Creator.

Number one and two, or first and second days, mean God in action; the Creator creating, or eternal being and doing. The meaning of one is God --- All in All, and the meaning of two, is the sum total of God's action.

This "Firmament" or expression in the midst of the waters is everywhere present. It is the heavenly work of dividing the waters from the waters --- producing the elements from the inherencies, preparatory to the producing of form, it is the Law of Expression in action.

31

"God divided" stands for the truth that God classifies all things and holds all things in their true relation to each other and in their true relation to Him, and knows Himself to be the whole.

"I am that I am; I will be what I will be. He who was, is, and evermore shall be."

Gen. 1:7
"And God made the firmament, and divided the waters which were under the firmament from the waters which were above the firmament; and it was so."

Ps. 148:4-5
The waters above the firmament are the potentialities of Creation. --- God's idea, or creation, before it is expressed. They are the possibilities of eternal life.

The waters below the firmament are all that results from God's possibilities called universal force, ether or elements, and what are formed within and of them. "God divided means God acted, and "it was so."

Gen. 1: 8
"And God called the firmament heaven. And the evening and the morning were the second day."

John 15:26-27
:"God called," is God proceeding forth as. He called the firmament heaven by being present with or in it; hence, it is firmly established. God is the dwelling place and origin of all creative action and creation. His calling is His issuing or coming forth. What he issues or brings forth is Himself made manifest, and this is the relationship

32

---

of heaven. The "firmament" is God's will being done. So God names it heaven, because it is so; "it is eternal, and the evening was and the morning was."

Return to the Top  Return to the Top

C. "Spirit of God" is comprehension; "moving upon the waters." - p. 33

GENESIS.

C. "Spirit of God" is comprehension; "moving upon the waters," exercising Its consciousness of Truth.

"The Spirit of God moved upon the face of the waters." Here we find the waters in the very foundation of all existence. "With thee is the Fountain of Life."

We find a marvelous statement of the waters as a foundation. "He layeth the beams of his chambers in the waters." The beam is the principle horizontal timber in a building --- this support must be the strongest and most durable that can be had.   Ps. 104 3

The Great Architect of the Universe finds in the "Waters" the stay and support of his buildings. True it is that our possibilities are all we have to build upon --- and emerge when we know these possibilities to be without limit.

"Praise Him ye waters that be above the Heavens."

"That stretcheth out the Earth above the Waters."   Ps. 148 4

"He leadeth me beside the still Waters."   Ps. 136:5

"My root was spread out by the Waters."   Ps. 23

Job 29: 19

33

Eccl. 11 1   "Cast thy bread upon the Waters, for thou shalt find it after many days."

Isa. 55: 1   "Ho! Every one that thirsteth, come ye to the Waters."

This is our possibility --- by the acceptance of such invitations, by the letting be of our Possibilities we consider what limits us to be our unbelief. The "Waters" flow out. In the outflow of the Waters we have rivers and streams. Divine Possibilities expressing. Waters represent Stillness --- Power in Repose. When I am led by the still Waters, my soul --- River --- is restored. Outflowing is action. River represents the outflow of our Possibilities --- Power being recognized in the individual soul.

Joel 3: 18   Rivers have their source in the Waters, are supplied by the Waters, are the Waters pressed out into action.

Ps. 46: 4   The Rivers of Judah --- Spiritual Consciousness --- shall flow with Waters."

Isa 66: 12   "There is a River, the Streams of which shall make glad the city of our God."

Ps. 65: 9   "I will extend peace to her like a River."

Ps. 78 16   "Thou visitest the Earth and waterest it; Thou greatly enricheth it with the River of God which is full of water."

Isa. 43:19-20   "He brought Streams out of the rock and caused Waters to run down like Rivers."

"I will make Rivers in the desert."

34

"There shall be upon every high mountain and upon every high hill, Rivers and Streams of Water. Moreover the light of the moon shall be as the light of the sun, and the light of the sun shall be sevenfold, as the light of the seven days; in the day that the Lord bindeth up the breach of His people and healeth the stroke of their wound."   Isa. 30:25: 26

"And he showed me a pure River of Water of Life --- clear as crystal, proceeding out of the throne of God, and on either side was the tree of Life and the leaves were for the healing of the nations, and they shall see his face and there shall be no night there. Behold, I   Rev. 22: 1-7 come quickly."   2 Kings 5: 10

The words to the leper were, " Go wash seven times in Jordan, and thy flesh shall come again to thee and thou shalt be clean."

John baptized in the river Jordan, preparing the way of the coming of Christ. We cross the river Jordan to reach Canaan. We must be purified ourselves of every wrong conception before we reach the consciousness of Perfection.

"Council in the heart of man is like deep water but a man of understanding will draw it out."   Prov. 20: 5

"Whosoever drinketh of the water that I shall give him, shall never thirst it shall be in him   Rev. 22:17
a well of water springing up into everlasting Life," Jesus told the Samaritan woman.

"And let him that is athirst, come and whosoever

35

---

will, let him take of the water of life freely."

Eccl. 1: 7
"All Rivers run into the sea and yet the sea is not full; unto the place from whence the Rivers came, thither they return again."

The "seas" are the result of the overflowing rivers; the sea represents the visible form that is set in bounds.

While we are in the knowledge of form alone, we are inclined to stop in form; we think we could be satisfied with the outer, but this is only personal belief, to stop in it would be stagnation. Understanding is the new birth that brings death to this misconception. In it the Waters bring forth abundantly and limitless consciousness of Life dawns.

Isa. 11 9
"The knowledge of the Lord shall cover the earth as the Waters cover the sea."

Rev. 20: 13
As consciousness of our Infinite Possibilities covers, obliterates, our sense of partial power. "The sea gave up the dead that were in it."

Hab. 2 4
"For the earth shall be filled with the knowledge of the glory of God as the Waters cover the sea."

Opinions and beliefs cause all sense of limitations.

Job 33:30
"He bindeth up the Waters in his thick clouds.""He hath compassed the Water with bounds

36

---

until the day and night come to an end." "The Waters are hid as with a stone and the face of the deep is frozen."

"The Waters wear the stones; these washeth away the things that grow out of the dust of the earth and thou destroyest the hope of man."   Job 14: 19

"My people have committed two evils; they have forsaken me, the fountain of living   Jer. 2:13
waters, .(the I Am,) and hewed them out cisterns that can hold no water."

Prov. 5:15-17
"Drink water out of thine own cistern and running waters out of thine own well. Let them only be thine own and not strangers with thee." The outflow is our possibilities is just as full and free as we let it be.

"Except a man be born of Water and the Spirit, he cannot enter into the Kingdom of God." The sick at the pool of Siloam waited for the moving of the Waters, and the first one that stepped in was healed. Jesus' first miracle was turning water into wine. Jesus walked upon the Water. Like Peter we may say, "Lord, bid me come to thee on the Water."

John 3: 5

Matt. 14: 28

We come to the Self, the consciousness of our sonship, only as we recognize, accept and manifest its limitless possibilities.

"Whosoever will, let him take the Water of Life freely." The invitation is boundless, the Waters are free. As we see this clearly, we shall

37

---

no longer be restricted to rivers and channels, but shall receive the flood of Infinite Power and Possibility that will make manifest all that I Am!

Return to the Top  Return to the Top

D. Thus His will is done in earth as it is done in heaven. - p. 38

GENESIS.

D. We must be in sympathy with, or in understanding of, the soul of things before there is light upon them. The evening of the second day means the mingling of God and heaven with earth, and the morning is the birth of the light of this truth in earth as it is in heaven. Thus His will is done in earth as it is done in heaven.

Gen. 1: 9

"And God said let the waters under the heaven be gathered together into one place, and let the dryland appear; and it was so." Simultaneous with the birth of light --- pure intelligence of God through heaven by creative action, the waters under the heavens are gathered together into one place and dry land appears. When God gathers the waters together, the elements of His manifest substance into one place, there is form. And it is eternally so; it is never otherwise. This is the first mention of the appearance of form.

Gen. 1: 10

"And God called the dry land earth; and the gathering together of the waters called He seas: and God saw that it was good." To call the dry land " earth, " is to come forth and be present within the earth; and to call the gathering together of the waters "seas," is to be in the act of forming

38

---

dry land; for the act of gathering together atoms of substance into one place called He seas, which is the act of forming, producing form. For God to divide the waters is for Him to be both the creator and creation; the omnipresent Spirit and its manifestation. Since all possibility of creation is contained in God and is the Waters above the firmament, so all the demonstration of His possibility in form is apparent in the Waters, or elements, below the firmament. God's eternal activity is symbolized in the literal sea, and the result of His action is actualized everywhere in literal form. As two, or second, means to do (Activity --- Expansion), the second day is for all, at all times; of what God does, and the third (forms -- result of expansion) day stands for all, at all times, of that which is done, And the whole is seen to be good.

"And God said let the earth bring forth grass, the herb yielding seed, and the fruit tree yielding fruit after his kind, whose seed is in itself upon the earth, and it was so." In the   Gen. 1: 11 potential earth, which is the idea of God, is contained the possibility of everything that is rooted in the earth. Being rooted in the earth symbolizes and is the expression of the truth that its reality is eternally in God. Fruit after His kind whose seed is in itself upon the earth, is God present and expressed in every kind. That which is born of God remains

2a

39

---

with It, Its seed, because It is born of God. As every tree is known by its own fruit, so every existing thing is truly known by its being the expression of one living and true God.

Gen. 1: 12

"And the earth brought forth grass and herb, yielding seed, after its kind, and the tree yielding fruit, whose seed, was in itself, after his kind; and God saw that it was good." The idea and potential earth brings forth all its possibilities after its kind. God manifesting Himself is God creating all, all creation is Himself revealed. This is good. After his kind means according to his kind in God.

Gen. 1: 13

Acts 7: 45-50

Matt. 5:34-36

"And the evening and the morning were the third day." All cause must show forth in effect, the Creator must appear in the Creation. When the Holy One acts, something is accomplished; when the One God creates, creation is the result. All result is form and all form appears in the third day. Three or third means rule --- result. The rule by which everything is done may be thus expressed: first, I am; second, I create; third, I produce; or, I am, I think; I speak. "I," is pronounced with limitless meaning. Creation is the spoken word of God, the expression of Himself, the revealment of what He is. "The word made flesh." (Appearance in what is called the physical.) As heaven is said to be the arch which overhangs the earth, the sky and atmosphere, the

40

---

place where the sun, moon and stars appear, so is it to be seen that God, the Infinite Spirit or Mind, in creative action overhangs and includes the earth and every form expressed, for they appear within and by His own heavenly act of self revealment. In His Divine activity, the sun, moon and stars are apparent. The same process is ever going on within each demonstration of Truth, and with the appearance of every living thing. When it is evening to one expression of truth, it is the blending of it into the morning or the birth of another expression.

The third day completes the detail of work contained in the statement of the first verse: "God created the heaven and the earth."

"The activity of Infinite Mind or Spirit is nowhere but in Mind or Spirit and moving upon the face of the waters, takes place within Itself. When we divide the waters from the waters, or discriminate between Cause and Effect, Creator and Creation, we are not to conclude that we have a lower and a higher nature that are warring against each other, two natures unlike each other. God made the seas of sense, termed materiality, of spiritual reality that they might gather together in one place, and produce form, and that the moving of the waters below the firmament might be known as the action of the Spirit of God in form."

"In the interpretation of the 11th verse of

John 5: 43-44

41

Genesis, Chapter I, it is essential that the law of expression be understood and applied, to avoid misconception. Most interpreters of Genesis at this point take up the dual doctrine and reason as if the earth was one source of living things, and the Creator another, which is a departure from wholeness. They forget that the earth is in God in the beginning and the grass and the fruit tree, and all

Matt. 6: 10

Matt. 6: 27

"The word that is with God in the beginning when spoken or expressed is good. Now we have the law of expression fulfilled in visibility. His done in earth as it is done in heaven. God, the ground substance of all existence. His action in the living visible result, whose seed is in Itself who is Itself the word --- is a trinity in unity (Creator, Creating and Creation), and is God's kingdom which is ever at hand in which there is naught but God.

The nature of all within it is Heavenly harmony. Heaven is within me because I have being in God. Earth is within me because my form or body is within God, and as one form is brought forth or caused to appear, so are all forms brought forth, and are made apparent. Strictly speaking, the Hebrew language has no tense system. Its verb form denotes state or condition, rather than time. Thus each Hebrew root serves to portray some method or degree, reached

42

in the action of the Divine Power. These primitive root significations, whose value hitherto has been so little known, enables the mind to rise from the world of form and phenomena to behold the glory and wisdom of God in Life and Law.

Truth is changeless, therefore in giving up our conceptions we are yielding our partial seeing to Truth. Truth is unchangeable here and now.

The deep is God's nature yet unfathomed or unknown. While man looks only on the surface, he does not know the depths of truth.

Job 38: 30

2 Cor. 4: 6

Return to the Top   Return to the Top

E. God bringing forth in heaven and earth. - p. 43

GENESIS.

E. God bringing forth in heaven and earth.

The individual does the work of the Father. Distinction between light and darkness. The glory of God declared on the earth. Signs, the formulated truth of Being. Days and years, the eternity of time. Inseparable individualities in heaven. Giving light upon the earth. I Am, the two great lights; the stars also. God, the nature and value of Creation. The use of law, not servitude, but fulfillment. Heaven and earth conjoined in the rule of the day and the rule of the night. The reality of body not seen by night but by day. The fourth day universality of the third. The heavenly host come forth in visibility. They give light upon the earth. The fifth day. The unity of waters above and below the firmament

43

The moving power of soul. The creature that hath life. Where life is. The fifth day universalizes the second. Living creatures free to move in heaven. All have right to live unto God. God's blessing. His presence with. Be fruitful, God covenant. No separation between God's creatures in substance and soul. His creatures partake of the nature of finished work. One differs not from another in glory, but to observation.

"I and my Father are One." When "I" is pronounced with limitless meaning, it stands for the same as God, the Father, whose nature is One. Unity forever is the state or nature of One. This unity is knowledge and power and is everywhere present.

Form is evidence of invisible Being, proof of the priority of invisible man--eternal self-hood; hence it is self-evident proof of the truth that I am. It is the word made manifest in the beginning, which is with God and is God. Form means, "And it is so."

Gen. 1: 14
"And God said, Let there be lights in the firmament of heaven to divide between the day and between the night; and let them be for signs and seasons, and for days and for years." The lights in the firmament of heaven are the conscious centers of Action within the Holy Spirit or Mind --- the individualities of the universal supreme Being --- the "heavens of Heaven."

44

"The firmament of the firmament," are God's idea of Himself, the potentiality of the Creator, expressed within and unto Himself and filled with the light of eternal glory. It is now clear why they are to do as God does; God divides the light from the darkness in the beginning and the waters which are above the firmament from the waters which are below the firmament and it is so now; hence our work is that of doing as our Father does.
Jer. 23:24
Deut. 4: 14-19

Individuality divides the day from the night and we are truly stars in the firmament of Heaven, distinguishing between the light of pure intelligence and the darkness of observation --- merely looking at and forming conclusions about things.
Ps. 74 16

Ps. 136:

True individuality declares the glory of God, for it constitutes the firmament, the totality of God in action, His complete method; and "as the firmament showeth His handiwork," so do we show signs, in formulated expression of the Truth of what we are; and we have seasons in which the innate power and possibility is brought forth; and the days and the years indicate the eternity of time in. which is expressed our eternal self-hood.
5-7

2 Cor. 5:

1-2

The days and the years blend all into one at the same time and that time is now, the acceptable day and year of our Lord; they stand for continuous light and eternity of the now in which, "The tree
Rev. 22 2
Eph. 3 15

45

---

of life bears twelve manner of fruits, and yields her fruit every month."

Gen. 1: 15	"And let them be for lights in the firmament of the heaven to give light upon the earth"; and it was so. Individualities are the lights in the firmament, which is eternally resting in God, and their state is the heavenly harmony of His own idea. They are to give light upon the earth --- not in heaven, for they are the idea of light eternally in the firmament of heaven, and they all exist in the "Son in the bosom of the Father."
Gen. 1: 16	"And God made two great lights, the greater light for the rule of the day, and the lesser light for the rule of the night. He made the stars also."
Ps. 148: 2-5	The "greater light" is knowledge innate, pure intelligence. Its rule is the activity of the Creator's innate idea of heaven and earth; the expressing of the word that is with God in the beginning, which is "the true light that lighteth every man that comes into the world."
John 3: 29-36	The "lesser light" is the expression, the intellect distributive with all its capabilities of knowing and reasoning according to truth, classifying and discriminating and letting light shine out in all our ways the same is the rule of the night -- what is looked upon with the natural eye. The lesser light then is mentality with its faculties illumined with innate knowledge, the greater light. So the truth of innate knowledge and pure

46

---

intelligence flows out through the firmament, or individual Divine mentality, into visibility.

The greater light stands for the self-illumination expressed by the Christ in the words of living light, "I and my Father are one" --- the true intelligence of God illuminating creation.	John 9:5
The lesser light stands for the knowing capacity of the expressed creation, pure intellect in which knowledge of self is expressed. "I am before I am manifested," is a statement of innate knowledge and the manifestation of that I am in the light that I let shine.	Rev. 22: 5 Ps. 8 3-4

"I am the light of the world," means I am eternal intelligence, and what I am is made manifest in the world of living things.

To rule the day is to Be the eternal life which lets shine.

To rule the night is to Be that perfect expression of life in everything that is looked upon. Seeing that innate knowledge is expressed in every part of the body, and in every form, is both the rule of the day and the rule of the night; this is true use of reason and judgment, and is evidence "Of my body broken for you."

The two great lights hold the same relation to each other as do Infinite Mind and mentality, as do God and heaven; or the universal and the individual; the same relationship as do the first day and the second day.	Ps. 104: 19

47

"He made the stars also." He made all the mental faculties and all divine thoughts. Divine Mind makes everything within and of Itself; not by nor through anything else but within and unto Itself exists the whole of creation; and all activity of His idea is set in the firmament of heaven to give light upon the earth.

Gen. 1:17	"And God set them in the firmament of heaven to give light upon earth."
John 6: 58	All of God's idea is expressed in heaven and heaven is to declare the Light of His glory in earth. The firmament is to show or outpicture His handiwork; each expression of the one Infinite Mind is in perfect harmony with every other expression. Their relationship to each other is found in the relation they each hold to the one Mind. As all examples in the science of numbers are seen to be in perfect harmony with each other because they are all expressions of one principle, and the value of each example is in the principle, so the nature and true value of each living expression of God is in

God. In this light the soul sees itself and realizes that the value of a man is that he has no duplicate

Eph. 4: 13

The stars or mental faculties are set in the firmament of heaven because they conceive, recognize and acknowledge the works of God. By them is understood and imaged forth the true purpose of the law of God, which is declared in the words,

48

"That having been begotten of Spirit is Spirit." The use of God's law is not that of servitude, but it is freedom of fulfillment and self-revelation. To give light on the earth is to see as God sees.

"And to rule over the day and over the night and to divide the light from the darkness; and God saw that it was good." To rule over the day and over the night is to show forth innate and intuitive knowledge or intelligence, and to manifest eternal power in mentality and visible form; and to know that heaven and earth are conjoined in God, and the whole is God manifest. To divide the light from the darkness is to know that "I Am" is the expressor of light and in Me is no darkness. Also that the expressions of God are apparent as such only in the light of eternal day, the intelligence of eternal life; and their real nature is never apparent --- not truly understood except in God's own knowledge of what He is.

Gen. 1:18

Jer. 31:35-37

John 7: 24

The truth of form is not seen by night, or in darkness, or from mere observation. The truth of what creation is, is the manifest or self-evident truth of what God is.

Rev. 21: 23

"And the evening and the morning were the fourth day."

Gen. 1:19

As one visible form is imaged forth in the firmament of heaven, so are all forms brought forth; and as the third day represents the creative act of producing form, "The gathering together

49

---

of the waters into one place," and the appearance is dry land, the fourth day is the same truth universalized, which proves that everything exists by virtue of the same law.

John 12:28, 45-46

John 1:14

Everything is made by and within the nature of one God --- Mind---to appear in form, in the face or open firmament of heaven. This is giving light upon the earth. The whole earth is made apparent or is understood within the light of heaven, where God's will is being done continually.

All innate power is expressed according to God's idea of Himself and through the individual, or indivisible mentality. The evening of the fourth day is the mingling and blending of God's heavenly hosts with all visibility; and the morning of that day is the coming forth in earth of all that constitutes heaven. All truth spoken or expressed is formulated Truth, and comes forth and appears in form and is the "Word made flesh."

Gen. 1: 20

"And God said, let the waters bring forth abundantly the moving creature that hath life, (Hebrew, soul) and fowl that may fly above the earth in the open firmament of heaven."

Rev. 22:1

Number five, or fifth, stands for law; its spiritual interpretation is "to array in order." It is the orderly method of bringing forth abundantly the moving creatures that hath soul. This command of God is to the waters both above and below the firmament which is in their midst.

50

The waters are creative waters of life in which the living creatures are made manifest. God creates every living creature which the waters of life bring forth after their kind. The waters above the firmament are God's power and possibility and the waters below the firmament are the possibilities expressed, and the firmament is the center of action in the midst, where the expression takes place, or between the expresser and expression, and in which the two are one.- This is heaven; they all have life in God.

The firmament of the second day testifies of the Truth that God, through demonstrating His own law --- love, brings forth one form and as he produces one, so are we all begotten and made apparent, according to His will and idea. The fifth day describes in detail the second day's work and universalizes the nature and result of it in visible form.

"Let fowl fly in the face of the firmament of heaven," stands for the truth that all visibility constitutes the face of the firmament of heaven, and that living ideas are free to move in   Gen. 7 14-15 heaven or fly above the earth. Moving creatures stand for the truth that their power and freedom of movement is God's breath of life within them. It is to be understood that the universal cosmic light is filled with fowl --- creative ideas --- that multiply in the earth. These ideas bring forth

51

---

everything after their kind; and we must see their divine right to life, liberty and happiness.

Gen. 1:21   "And God created great whales and every living creature that moveth; which the waters brought forth abundantly, and after their kind, and every winged fowl after his kind; and God saw that it was good."

Gen. 1: 22

"And God blessed them, saying, 'Be fruitful and multiply, and fill the waters in the seas, and let fowl multiply in the earth.' "

Return to the Top   Return to the Top

F. "And the evening and the morning were the fifth day." - p. 52

GENESIS.

F. "And the evening and the morning were the fifth day."

Gen 1:23   For the waters to bring forth abundantly the living things that bath soul, is for God to fulfill His own law of expression. The fact that all these things live and move now is proof of their simultaneous expression and of their beginning of existence in God at this time; and argues the eternity of creative action, the ceaseless activity of God.

"God blessed them, saying, 'Be fruitful and multiply.'"

Gen. 9:12-5

His blessing upon them is His presence manifest in them, it is at-one-ment of Creator and Creation. God keeping the covenant with them for perpetual generation is the saying, "Be fruitful."

52

All forms, colors and qualities are inherent in the one God and emanate from Him and are good.

The evening and the morning of the fifth day testify of the union of all living creatures with the Creator and the mingling and blending of all living and moving things with all that is described in the sixth day. These living creatures are distinct expressions, separate in shape or outline, but in no way separated in life, soul or idea, or by quality of substance; all are one substance. For them to be fruitful, they must necessarily have being in the seed or word that is with God in the beginning, without which is not, anything made that is made.

Rom. 8:21-23

1 Cor. 3:6

The infinite variety of expression is one in Being, and the whole is good. And the morning of the fifth day is the coming forth of God within and through all things that He has created --- expressing His image and likeness in the whole. One thing cannot differ from another except to observation. The whole is one eternal good.

Blessing is bestowed upon living, moving creatures because God is Love and Love's idea of co-operation is manifest in them. Blessing stands for the state of being blessed; they are blessed with God life and power, which is the covenant for perpetual generation.

"And God said, 'Let the earth bring forth the

53

Gen. 124

---

living creature after his kind, cattle and creeping thing, and beast of the earth after his kind'; and it was so."

Gen 1:25

"And God made the beast of the earth after his kind, and cattle after their kind, and everything that creepeth upon the earth after his kind; and God saw that it was good."

To see God, is to perceive His law fulfilled, expressed in creation. To conform to His law is to conform to the truth that it is already fulfilled in creation, expressed in the living. This understanding and obedience is power and satisfaction.

Gen 8:17

As firmament or heaven is in the midst of the waters, so God's idea of heaven contains His idea of earth; hence it is to be remembered, that where the earth brings forth, instead of the waters, His idea is being imaged in form, which is His possibilities, revealed in heaven, exactly like unto Himself in nature; hence all that is potential within is expressed in the infinite variety of forms that are apparent, and these forms are eternally in God. They are all expressions of His idea of creation; the product of Divine Creative principle.

Matt. 11: 25-28

Luke 10:19

Luke 18:17

There are no creatures but what God makes and they are all good in their nature. "After their kind" is according to their origin, and as their origin is in God, who created them, their kind is according to God's idea of them. All

54

---

creatures are in a state of harmony and loving unity with each other and God sees that the perfection and the harmony of the whole is good. There are no carnivorous creatures begotten of God. Their true nature is portrayed by Isaiah, for in God's kingdom where the All in All is God and God manifest; the wolf lies down with the lamb; the leopard lies down with the kid; the calf and the young lion and the fatling are together; and a little child leads them. Unless we see all to be as pure and innocent in their nature --- which is God given --- as are little children, we shall not be, in our perception, as little children, and shall not realize that we are in the kingdom of heaven.

"And God said, ` Let us make man in our image, after our likeness; and let them have dominion over the fish of the sea, and over the fowl of the air, and over the cattle and over all the earth, and over every creeping thing, that creepeth upon the earth.' "   Gen. 1:26

Gen. 1:27

"So God created man in His own image, and in the image of God created He him; male and female created He them."

"And God blessed them, and God said unto them, 'Be fruitful and multiply and replenish the earth, and subdue it; and have dominion over the fish of the sea, and over the fowl of the air, and   Gen. 1:28

55

---

over every living thing that moveth upon the earth.' "

The Hebrew root meaning of six or sixth, is to set, fix, establish, complete.

Matt. 5: 1 "Elohim" is a plural term, translated God. "Let us," means "we do" make man in our own image. This plurality of Spirit does not imply more than one God, but refers to God and His idea, man; the nature of this idea is indicated in divine communion, co-operation and oneness of purpose of God and man.

Gen. 5: 1-2

Number six denotes the finished work. Man is the family name for humanity, for the sons and daughters of God. The image of God Himself imaged, expressed in form. " The word became flesh, and dwelt among us." This image is exactly like God in all that He is. This is the only man, male or female, that God ever made and there is none other.

Phil. 2: 5-7 "Let them have dominion" means let them be conscious of being life and lord of all things, as I Am. Let them see themselves expressed in the whole. Let them be what I am, exist as I do, and be --- have--the same dominion that I am. God dominion is that given to man, male and female.
The whole is to be seen as God and God manifest.

John 5: 14 The work of man is not that of rising from a lower to a higher plane of Being, or of unfolding into Divine dominion. His work is to do what he

Eccl. 7: 29

56

---

sees the Father doing. God expresses His identity in man, and man is to see God's omnipresence expressed in all. Man, God's image and likeness, is potential in God before he is expressed, and he is therefore co-worker and equal with God, capable of doing His will. There is no gulf to be bridged between God and any of His creatures in heaven or in earth.   Phil. 2 5,6

"And God said, 'Behold, I have given you every herb bearing seed, which is upon the face of all the earth, and every tree, in which is the fruit of a tree yielding seed; to you it shall be for meat.'"   Gen. 1:29

Gen. 1: 31

"And to every beast of the earth, and to every fowl of the air, and to everything that ereepeth upon the earth wherein then• i'. lire, I have given every green herb for meat; and it was so."

"And God saw everything that He made, and behold it was very good. And the evening and the morning were the sixth day."

As in the third day, the plane where form was finished work appears, God brings forth the fruit tree yielding fruit after his kind whose seed is in itself upon the earth, so the further expansion of this idea is to be found in the sixth day, which day describes the nature of the whole and makes known the true use of things.

We find that God gives to the individual for food the comprehension of the truth of Being;

Matt. 4-4

57

---

the creature is sustained and fed in that it partakes of the nature of the Creator. "Man does not live by bread alone, but by every word that proceedeth out of the mouth of God."

John 17 4

"And God saw everything that He had made, and behold it was very good." A good God and Father whose nature is love loves all things into existence. There was never a time when He did not make everything that was made and see everything that He made, and pronounced upon it the blessing, "It is very good." Heaven and earth are both spiritual creations, the fulfillment of the law of expression.

The Sixth Day being a full description of the nature of the first, and the first including all in the sixth, we perceive that if in the beginning God made all that was made, there is never a time when He does not make all that is made. The fulness of rest is I Am All and in All.

Since the starting point of creation is God and since He is His own creative activity and result, the completion, the seventh day, finds us, being and existence in God. This is rest --- the complete perfect at-one-ment. All is good and very good now.

Why all is good.

It is God who creates, who moves upon the face of the waters, who says, "Let there be light";

58

---

who sees light as good, who divides it from darkness and calls it day.

It is God who says, "Let be a firmament," who divides the waters from the waters, and calls the firmament heaven.

It is God who says, "Let the dry land appear," who calls its appearance earth, who says, "Let the earth bring forth," and sees it all good. It is God who says, "Let be" lights in the firmament of heaven, let them divide the day from the night --- light from darkness --- and give light upon the earth; (i. e., Do as I do.)

It is God who is the two great lights --- Being to rule the light --- existence to rule the darkness over the face; and sees them good.

It is God who creates every living creature that moveth, which the waters bring forth abundantly; who blesses them and says, "Be fruitful."

It is God who makes the beast of the earth and everything after its kind, and sees all good. So it is God who creates man, male and female; who blesses them, by being present in them; who commands them to replenish the earth and have dominion.

It is God who gives man food, and who gives food to the beasts, fowls and creeping things wherein there is a living soul.

59

It is God who sees everything that is made, and who says, "It is good and very good."

It is God who finishes his work by pronouncing it good and who rests in its goodness.

Return to the Top   Return to the Top

G. Allegory. - p. 60

GENESIS.

G. Allegory means the description of one thing under the name of another a figurative sentence or narrative in which the principal subject is kept from view, and we are left to perceive the meaning of the writer by the resemblance of the picture drawn to the primary subject.

Isa. 64: 8

The watering of the whole face of the ground is the bringing forth of all of God's possibilities into visible form. " The Lord God formed man of the dust of the ground," means that He formed him of the atoms of Infinite substance, Ground of Eternal Being. Forming man is the fulfilling of the injunction, "Be fruitful and multiply and replenish the earth, and have dominion," --- that is, be the life and substance of.

So the figures of the subject we are considering is true to the Divine record of creation, and it is the Lord God who breathes into his nostrils the breath of life a direct out-breathing of the Holy Spirit, and the result --- "And man became a living soul," a conscious individuality.

Gen. 2: 8-10

In the 8th, 9th and 10th verses we have an allegory, a picture of the whole of creation portrayed

60

---

as the garden of God. Eden is the ground in which the Lord God plants the garden and there is where he puts the man he forms. Eden is the ground of eternal substance; it is the dust of this ground of which man was formed. There is but one substance --- this is Holy ground, and for this reason, man was named Adam in the day in which he was created, the sixth day of finished work.

The Hebrew word used for garden, gan, denotes that which has been covered, protected, that fruit be raised. To plant, is to set, fix --- the image and likeness is set, fixed, in its Source, eternal Being.

"Eastward" means where the light is. God is light and is the cause of all that comes forth and all things consist in light which He calls day. Therefore, the garden planted eastward in Eden, necessarily stands for the truth that all creation, invisible or visible, originates and lives in God.

Col. 1: 16-17

John 1: 1-2

John 17: 5

Eden --- Adhen, denotes pleasure, delight; it stands for potential creation, the garden planted. The innate possibility of the creator, a garden where everything is rooted in the ground allegorically expresses the truth that the whole creation is the image of the invisible God. The whole has being and origin in God; we are all firmly rooted and grounded in Him, as is a literal garden planted in the ground. And we can no more bear

Acts 17: 23-24

Col. 1:13-16

61

Rom. 8: 29

---

fruit apart from God, the Infinite Source, than can a branch severed from the vine.

Acts 17:24-29

And the tree of Life in the midst of the garden stands for the truth that one life is in the midst of every living thing; that everything planted in Eden has being in God and that everything that lives is in harmonious relation with every other thing, and all are at-one with God.

We live in a realm of law and order. The knowledge of the tree of life is the knowledge of this Truth, that the whole is God, and also the knowledge lint to depart from this truth of the omnipresence of God is evil or ignorance. This wonderful truth has been set forth in this allegory that we might look to the primal subject, our basis, God and His expression. When we do this, we do not judge from observation.

Rev. 20:15

Isa. 37: 16

The written testimony of the truth of God's garden planted eastward in Eden is one of Life, and will forever remain a mystery to those who have not learned the truth of Being and the Law of Expression.

Eden stands for the whole of God, potentiality and power, in the light of intelligence, the pure knowledge that illumines all; the knowledge that knows what is to be, now; knows that there is no power in opposition to Being. The tree of knowledge stands side by side with the tree of Life in the midst of the garden; its office is Judgment of

62

---

truth according to the Infinitude of God. It must ever be the flaming sword --- the sword of Truth.-- which turns every way, discriminates always and everywhere, to keep the way of the tree of Life.

The tree of Life in the midst of the garden of Eden is unity of the Creator with heaven and earth, the paradise of God.

Rev. 2:7

Forbidden fruit --- miscalculation of values. There is only one knowledge, so the figurative representation of knowledge as a tree, is called "The tree of knowledge of good and evil." The perfect knowledge of God discriminates between perfection and the false claim that there is imperfection; between being wisdom and being made wise through experience; between the true state of the finished works of God and the theories that they are to be finished later through personal effort; between being whole or Holy and the many theories by which we are to become whole by what we do.

Luke 11:9-10

2 Sam. 22-23

"And the river went out of Eden to water the Garden and from thence it was parted, and became into four heads."

Gen. 2 10

Since the garden of Eden stands for the whole of God's creation expressed, it follows that all related of it must describe different parts. This is shown by the etymology of the word "river --- Nahar," to flow, to move, to cause to shine. It is the outflowing of the Holy Spirit. God's glory,

3

63

---

the Light that God is shining in earth as the fulfillment of Law.

Like all other forms of water spoken of in the Hebrew scriptures, it denotes activity, creative power, a never-ending formative action, by which all forms of life are sustained and lived.

Four heads --- "four" means perfected existence of heaven and earth, of all action and result, and the four heads indicate the complete and perfect state in which all things are lived and sustained by one Supreme Being. In the human form it is symbolized in circulation, respiration, digestion and generation.

Under the symbol of a garden watered by the mists of the earth, we have a picture of soul development in which man has reached the consciousness of existence in the body, but does not understand the cause of existence, nor the source of his possibilities. The picture is beautiful in that it represents a soul fully equipped with all innate possibilities and manifesting beauty and goodness in outer form.

The garden was growing "every tree that was pleasant to the sight and good for food." In this stage of unfoldment, that is satisfactory which appeals to the senses, sight and taste are gratified.

The "face of the ground" is watered by a "mist that went up" from the earth because man

64

---

is as yet ignorant of cause and effect. He seeks satisfaction in things without, and considers them as the source of all good. His consciousness is centered in the body, the earth from which he believes his possibilities spring.

The "mist" well represents what can be received from this stage of consciousness. But the rising of the "mist" may typify man's aspiration after the higher, the prayers that have "gone up" from the place of undeveloped understanding.

The Garden of Eden represents the finished body.

Now, man's work must be of a higher nature; a complete change takes place. It comes about through fuller illumination, in which light the importance of the soul begins to be seen and for the time the body is humiliated.

This enlarged consciousness is indicated by the parting of the "river" in Eden into "four heads." The name of the first, Pison means "changed"; the second, Gihon means "Valley of grace" or illumination through inward light; the third "head," Hiddekel represents "Voice or sound," which means expression 'the step that must follow more light; and the fourth, Euphrates means "making fruitful." These well represent the result in manifestation that comes with expression.

Man cannot remain in a condition of ease from

65

---

innocence. The fully developed body well supplied apparently from the earth, with every sense gratified is a step to something beyond. Within are the soul qualities struggling for birth and that birth is understanding. By the inbreathing man becomes conscious of his soul, but faintly conscious at first.

Matt. 13: 25   The first "three days" stand for soul unfoldment and the second "three days" stand for the understanding of that enfoldment. The tree of the knowledge of good and evil is the light and the darkness that attend upon the way in enfoldment. Up to the day of understanding, the soul has no temptation to form opinions, to handle what it does not understand. While understanding is still young --- undeveloped--there is a strong temptation to express opinions, but Wisdom says, "Thou shalt not." Parable of wheat and tares.

The soul unfolded to its consciousness of form, or body, is the Garden of Eden; the place of being without understanding. The "living soul," created and put into the garden to dress it, is the enfoldment of individual consciousness to the place of understanding. The soul begins to see its responsibility and to assume care of the body. Man here enters the age of reason, begins to comprehend cause and effect. Soul is recognized as the essential nature; but in the process of unfoldment into consciousness of Truth body is sub-ordinated

66

---

to soul. The soul is to expand in power and control the body; this is What the soul is awaking to, but it is only the first step in understanding. The first step in unfoldment, body is held as below soul, as the servant of soul.

In fuller consciousness it is known as the helpmeet for the soul. Verse 18. And the Lord God said, "It is not good for man to be alone; I will make a helpmeet for him." 21 "And the Lord God caused a deep sleep to fall upon Adam and he slept, and he took one of his ribs and closed up the flesh instead thereof."

Gen. 2: 18

Gen. 2: 21

The Hebrew word "Ahdam --- Adam" is not merely the name of a person but, like the Greek Anthropos and the Latin Homo, it is the class name of a genus or group.

When God creates Man, male and female, he calls their name Adam, in the day (the same light) in which he created them.

Gen. 5: 1-2

The account of the creation of Eve from the side of Adam allegorically represents the truth that God creates Man, male and female, of one substance. "Female" stands for the generative power of God's idea and of the mental receptivity of Man universal, male and female. Through holy conception, the true mental state, are both visibly formed; thus the true mentality is the help-meet, Ahzar, which means that which girds, surrounds, defends. It is clear that it is not good

67

---

for man to be alone; and since male and female are not alone in God, in their Being, it is not good for them to be alone mentally or bodily; and just as the male is formed out of atoms of eternal substance, so the female must be formed, that man and woman may stand as one in Being and existence. Thus the declaration, "male and female created he them," is a spiritual declaration actualized bodily in truth; they are as the angels of heaven. As all law and prophecy were fulfilled in the Christ so the consciousness that God and God manifest is the all in all portrays the truth of Man, male and female, and their marriage in God. This is not only prophetical of what is to be established in the social relations of mankind, universally, in the coming age, but it is the truth always, and everywhere, waiting our recognition.

It stands for the truth that they are both the image and likeness of God.

Eph. 5:29-33   The conception of truth and the receptivity of man mentally and bodily, called the "rib," represents the truth that male and female stand side by side in form, and neither is to mentally rule over the other (nor to dominate the body of the other). Both, in Truth, are in the Holy Spirit bodily.

Since God is the only head of Creation, Eve could not have been taken from the head of man, nor could she have been taken from His feet,

68

---

since the earth is the Lord's and is His manifestation.

Man created male and female represents mankind. He formed Man of the dust of the ground, represents the forming of mankind. The Idea Man, male and female, must appear as two forms to fully express Man. Since Man the Idea contains both sexes creating is

Rom. 11:16

making woman out of Man, but no more than it is the making of man out of Man. Verily nothing can be expressed that is not before it is expressed. The reason that it is represented that she is taken from his side is that the only difference in them is sex, and they are side by side in the Lord-Law. So the understanding of truth must represent them as standing side by side, having one purpose, one conception, one Godhead.

"And closed up the flesh instead thereof," stands for the truth that woman is formed or budded without loss of substance to man; there is one substance manifest in both. It stands for the truth that Being is not diminished by its Infinite variety of manifestations, and that the only way is God's way; so Man, male and female, obeying the injunction, "Be fruitful," multiply and replenish the earth with form, cannot diminish power or substance.

Their expressions must come forth as themselves consciously made manifest. For Man

69

---

universal to see himself manifested individually in his offspring and they in Him, is to see them as "bone of my bone, and flesh of my flesh." Man, male and female, Adam and Eve, men and women, are one Spirit and one body. The body is Spirit Substance. " That which is born of Spirit, is Spirit."

"This is now bone of my bone and flesh of my flesh," means spiritually this is Being of my Being, and existence of my spirit of my Spirit, and Substance of my Substance.

Job 14: 1

Matt. 19: 4-7

This is the origin of marriage, the rock foundation upon which it rests. Therefore shall a man leave father and mother (the belief that they are the Source and Cause of his being and existence) and cleave to this eternal unity of Being and existence, that his Being is God and his existence is God manifest.

Rom. 7: 4

"And they were both naked" --- they were both as God is, just as He caused them to exist. "And they were not ashamed" --- they were not fearful, for there is nothing portrayed in this state and kingdom but God and His work, God in His relation to the whole of Creation.

Adam and Eve represent the two stages in developing consciousness called the personal and individual, reason and intuition, body and living soul, which must understand their unity before harmony is realized.

70

"And the rib which the Lord God had taken from Man, made He a woman, and brought her unto the man." Tsalah --- rib, denotes a side, an extension. First claim of separation of soul and body. There is but One. That One is Mind and it is Omnipresence, including every phase of development through which the soul passes on its way to full consciousness. Woman is a phase in this One; man also is a phase in this One.

Gen. 2: 22

Woman stands for intuition; man for the reasoning faculty. It is now seen that body is the offspring of soul; that body is of the very substance of soul; or as Adam speaks,

"This is now bone of my bone and flesh of my flesh; she shall be called woman because she was taken out of Man." As this is but the dawn of understanding, we need not wonder if the full relation of this "pair" is not comprehended. Only now do we begin to realize what the body is to the soul and what the soul is to the body. In generation, woman --- intuition--is the governing force; in regeneration, man --- reason, rules; in complete understanding there is recognition of unity; Body is seen to be as sacred as Soul; intuition as great as reason; woman is as important as man, because equality is realized and neither rules the other. This is union of understanding.

Gen. 2., 23

In generation there is no thought of the lifting up of the body. While the body governs us,

3a

71

---

it cannot be rightly understood or lifted up; in regeneration begins the thought of raising the body. The uplifting of woman and the uplifting of the body are simultaneous; both processes are at work today.

Man, in a certain stage of development, separated woman from himself. As a consequence, he has had to wait upon woman, wait for the full development of intuition before he could know how to reason. Now, in turn, intuition waits upon reason, follows and obeys, until the fulness of reason is come in understanding.

Neither is first. Woman acts in the love of wisdom; man acts in the power of understanding; the blending of the two produces Life in Perfection. Man sent woman forth from him; man must redeem woman, bring her back.

In the lifting up of woman, man may see the lifting up of his body. "And the twain shall be one flesh"; soul and body known as one. In the first step, the One seems corning forth into all; in the second step, all things return to the One; in the third step there is neither generation nor regeneration --- "Neither marrying nor giving in marriage"; no making of anything, but the realization is of the unity, that never was broken.

72

Return to the Top   Return to the Top

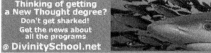

H. The dawn of understanding. - p. 73

GENESIS.

H. The dawn of understanding.

Why is Man entitled to this fuller unfoldment?

We never receive anything until we are entitled to it. Man can receive this new light, in which is to spring forth his conscious power, his understanding of limitless possibility, because he has done what he could. The mist going up and watering the face of things may not represent deep knowledge, but it does show man as using what knowledge he has. The church is working according to its knowledge; Christian Science according to its understanding; Divine Science according to its realization. The result in each case shall be more light. We cannot ignore a step that has been taken in our mental and spiritual progress any more than we can hold in contempt our first physical steps. Each has its place in our unfoldment, and that place is best understood after we have gone beyond it.

"And the Lord God planted a Garden east ward in Eden and there He put the man He had formed." "And out of the ground made the Lord, God to grow every tree that is pleasant to the sight and good for food. The tree of Life also in the midst of the Garden, and the tree of knowledge of good and evil."   Gen. 2: 8

The first "three days" represent our unfoldment into the consciousness of form.

73

Matt. 15:13	This is the "Garden of Eden" period. Implanted in the individual soul is everything that is to be expressed. A garden is a planted ground and in it things are growing. This soul garden is planted by its Master. The individual does not plant. God, the Father, is the planter; man individual is the keeper and cultivator.
Gen. 2:15	"And the Lord God took the man and puts him into the Garden of Eden to dress and keep it." The soul is the garden; thought does not make nor sow the "seed" --- these are the substance of Divine Mind that we now know and understand. Thought is to cultivate, exercise, dress, keep -- that is all.
Gen. 2: 16	"And the Lord God commanded the man, saying, 'of every tree of the garden thou mayest freely eat.'
Gen. 2: 17	But of the tree of knowledge of good and evil thou shalt not eat, for on the day thou eatest thereof thou shalt surely die.' " This tree was not in the garden.

We do not see the darkness as evil; it is ignorance. There is little realization of Light but all Light is there. The greater part of the Light is unseen, uncomprehended, it is darkness to us, but nevertheless it is Light. Man, from the first moment of dawning consciousness, is to look at the Light he does see! he is to work with the wisdom he has received; he is to use all he has

74

been able to grasp. Doing so, he will develop into fuller and fuller consciousness; the Light increasing and the "darkness" decreasing. He is not to concern himself with the "darkness"; he is not expected to handle that which he does not yet understand. He is to "let be" that which is unfolding within him.

Man is to use his knowledge, but he is not to form opinions about what he does not know. If he does, he will experience inharmonious conditions out of which he must work his way. This is partaking 'of the "tree of the knowledge of good and evil," a thought divided between certainty and uncertainty, a mixture of trust in what he knows, and doubt or fear of what lie does not know.

The light and darkness are one; the former is the extent of our knowledge; the latter is Truth unknown. "Evil" is the result of forming personal opinions about the unknown; it is the attitude of impatience, which destroys trust, breaks our calm and makes us restless. This is the cause of all ills, of all unhappy experiences.

Do we have to learn by experience?

There was only one day to God. Everything formed out of the ground means unfolded from the Source-Basis. Man is in the first day --- we shall know that every soul is in the Beginning. Some things in the Bible are written

Gen. 2: 4-5

Gen. 27, 9, 19

75

from the standpoint of process; others after the absolute.

All that is true, is true to Spirit or God. No chapter in the Bible bears a physical or personal meaning. Each has meaning in its relation to God. The present worldly view of form and visibility and all that is of it, shall cease, but this body which is the word of the Lord shall endure forever; it shall endure as Spirit and Life.

Rom. 8: 23, 39-39	There was and is a threefold purpose in the Gospel of Jesus. First, to show what God is; second, to show that matter (so-called) is not evil and that form or body is misused by the false beliefs held about it, hence, by the Gospel of Truth the sick are healed, the

dead are raised and God's will is done in earth as it is in heaven; third, to make known the means by which the misconception of body will cease and the Truth of the perfect body will be demonstrated. This redemption comes through at-one-ment. This consciousness is life, and life abundant, for it brings Eternal Life and immortality to light.

Phil. 2:6

Rev. 12: 9

Gal. 3: 20

The means by which we cease the misconception of the body is through knowing the truth of the being and the existence of man --- the conception and birth in equality in God. The serpent, tempter, is not matter, is not spirit, is not that which makes the true use of form, but that which would give form the precedence. Therefore, it is written,

76

"Now the serpent was more subtle than any beast of the field which the Lord God had made. And he said unto the woman, 'Yea, bath God said, Ye shall not eat of every tree of the garden' "? The serpent, being more subtle than any beast of the field which the Lord God had made, shows that He did not make it; if He had it would have been in harmony with, and not more subtle than any beast of the field. It has no place in Being, Substance or Reality. Seeing this truth casts out ignorance, the serpent, the accuser and deceiver. The spiritual interpretation of meat is that of which we partake spiritually, with which we spiritually commune, and with which we are in a state of oneness. To eat is to partake of, to assimilate and be one with.

We are not to believe that effect is cause; to do so is to believe that there are two causes --- one beside the great first Cause --- both working in us, and that the effort to serve the one must be made to escape the influence of the other. "Ye cannot serve God and mammon."

2 Cor. 16

Luke 16 13

The Lord God commanded man. saying, "Of every tree of the garden thou mayest freely eat." He spoke this truth to man, which means Being thou shalt Be, for beside Being there is nothing, so it means to be what God is, or not to Be at all; in other words, it is eating of what is, or starving. Everything being made in light, it is made in wisdom, and God, the maker of all things, is the light

Gen. 2 16

1 Thess. 5:23

77

and wisdom of all. So, of the tree of the knowledge of good and evil thou shalt not eat --- or be of it. "For in the day thou eatest thereof thou shalt surely die. In that light in which the claim is made that we are both good and evil, we are separated, in belief, from the Lord God. This is when the belief comes that "man is twain," that soul and body must he parted, and that death is the means by which we enter life.

Divine communion does not consist in contradiction; communion is at-one-ment, like communes with like. Man is to work as God works; he is to be the light and wisdom that illumines his own works; his works are not to give him knowledge or wisdom; man is. I am before I act.

1 Cor. 10:16-17

Luke 20:36-38

Nothing is created by creation, hence, no work is done but that which is the result of knowledge and wisdom. In truth the eye is single and the whole of creation is full of the light, knowledge and wisdom of its Maker. The appearance of the serpent is the appearance of a doubting question; and when we attempt to state the truth of the commands of the Lord God to it, as did Eve, it contradicts God's command, hence it is said, "Ye shall not surely die." Not knowing that dying to the truth that all is good, or God, is separation, and that separation is death to the truth of Holy Being --- being whole.

So the serpent says, "God doth know that in

78

the day ye eat thereof (believe that ye are good and evil), then your eyes shall be opened; and ye shall be as gods, knowing good and evil."

To be as gods is to be in belief separated from God and is to believe that there can be more than one God, and that we are separate one from another in our natures and in our relationship to Him; that we are not one in Spirit and body.

John 4 24

1 Cor. 6:9-10

"And when the woman saw that the tree was good for food," she evidently believed that it was good to partake of, hence good to experience contradictions; and when, she concluded that it was "pleasant to the eyes, and a tree to be desired to make one wise, she took of the fruit thereof." To conclude that anything in creation is to make creation wise is to ignore the truth that there is one Source and Cause of wisdom that makes all wise, and that all creations are always being created in the light of pure intelligence.

Eph. 4 4-6

Reasoning that we are to be made wise in the future, and denying the fact that we have been and are being made wise now, and that we now exist in the light of wisdom, belongs to the forbidden fruit. He who reasons as if effects --- works done --- are causes, the source of growth and enfoldment, is ever seeking and never finding; always expecting to be made wise, but never touching even the hem of the garment of wisdom.

Col. 1: 12-19

Observation does not make wise, but knowledge

79

and wisdom are the basis of intelligent observation.

To believe that the experiencing of opposite results within ourselves is a wise teacher is to mentally partake, and is to give also unto the husband or unto that to which we are wedded either in being or in seeming. From the moment that we in our thought change the Source of life, wisdom, health and supply to the external, we have, in belief, husband or wife separated from the Lord God. Thus the creature is worshipped and served more than the Creator. The visible body is then supposed to be separated from its invisible Source or Cause; it is no longer the wisdom of the Lord God manifest as Man, but it is men and women --- gods--gaining knowledge and wisdom through experience. This denies the truth that they are like God, his own image and likeness they hope to become like Him in another world, and to obtain heaven through death of the body.

Phil. 2: 5-7

Just as soon as opposites are claimed to be the Source of wisdom, "good to make wise," that claim separates the man visible from God invisible and ignores the relationship existing between the two. Just here is where belief in duality begins to clothe the outer man with fig leaves, raiment, the source of which is believed to be external and not God. This false conception

80

covers up the real spirituality of the body with the belief of materiality.

The tree, knowledge, itself knowing the all and all to be good and that evil falls short of the truth that all is good, is a guide to right action, thinking, and right word.

God is always right. His ways are just and good. To conclude that evil, ignorance, is good to make wise and that it is undeveloped good, is to ignore the fact that its only meaning is a missing of the truth that we are already made wise, and that, in the Lord God in the bosom of the Father, "I am the Son, the light of the world."

The serpent is the enmity to truth, contradiction of the Infinity of God; it is figment of belief, the supposition that there is another power acting in us beside the Lord God, or God Almighty: Knowledge knows that there is nothing more and nothing less than the Infinite and Omnipresent One; that in the Lord God, perfect Being and perfect doing, all exist in Light and are made wise.

There was nothing the Lord God made that was like the serpent in its nature; there was nothing in all that He made, like a query or doubting question or accusation; nothing like "If thou be Christ, save thyself and us." Luke 23 39

To realize the truth of form, one must be wise and knowing. We cannot be made wise by comparing

81

---

things with things supposing them to be opposites in their nature, reality and substance.

Rom. 8: 28
The Lord God teaches that like produces like, and that we cannot have two opposing results from one cause. "Out of the mouth of the most high cometh not forth good and evil." All things, results,, work together for good, and are good, and are so seen when their relationship to their Source is seen.

John 8 44
To know that God and God manifest is all is to know what Being is, and what I am, "thou shalt be this truth and be holy." Hence, Jesus said to the false beliefs, judgments front observation, which thought to kill him, "Ye are of your father, the devil, and the lusts of your father ye will do."

John 7 24
You shall not touch the forbidden fruit, nor desire to be made wise, for the only way to know wisdom is to know God, and to know Him is to abide within Him, and not doubt the judgment of truth. The beginning of miscalculation is subtle. It comes through giving attention to what is either impressed upon the retina of the eye or sensed from observation.

They thought that the effect of the tree of Divine discrimination would make them wise, instead of being the wisdom that uses the tree. One must Be wise to use Divine discrimination. An opinion drawn from observation without the use of the judgment of truth never gets away from

82

---

observation, and effect is its only source of wisdom; it is called the serpent because it shall go, or move only on the surface of things, and partake of or eat dust, which means it can only partake of appearance.

The command is that we shall not believe that we can have within ourselves from one Source, two results which are opposite to each other. Since there is but one source in Truth, all results are good, for a good Source and Cause cannot bring forth an evil result or effect. The eyes opened to good and evil, means they are opened to duality, opposites; they are ready to contrast appearances, but do not compare living things with God; so conclusions are ceaselessly being formed from what is observed.

Our thoughts, words and deeds are governed by these conclusions; and, since they have no basis but effect, they contradict real knowledge, hence there is apparent contradiction going on in thought, the inharmony of which words, deeds and sensations partake. Then it is that Adam hears the voice of the Lord God in the wind of the day, not within himself, and it is true at all times, with all people, that when they look without for God, they do not hear his voice within and have hid themselves in external things. So it is said that Adam and his wife hid themselves amongst the trees in the garden. This represents a belief in

83

---

separation, "And the Lord God called unto Adam, and said, 'Where art thou?' And Adam said, 'I heard thy voice in the garden, and I was afraid because I was naked.' "

Adam was afraid to appear in the presence of the Lord naked. This means he was afraid to appear as the Lord God made him, as God's own image and likeness. This condition in people comes from their having hid themselves in external things, "amongst the trees in the garden." The conversation which takes place between them and the Lord after this is as between two persons.

Heb. 8:9-10 To this recognition the Lord God no longer speaks in man but seems to speak to him from without; and He says to this serpent, or mental belief, "I will put enmity between thee and the woman and between thy seed and her seed, (or between thy word and her word); it shall bruise thy head, and thou shalt bruise his heel." Discord is sensed only in the heel or visible body or form, and it is sensed there only because we mentally listen to impressions made from observation, and let them voice themselves or take form in word and deed, and thus reject the consciousness or true conception and birth.

"The seed (or word) of the woman shall bruise thy head," means that her Word shall destroy the serpent; her word being truth, it shall destroy all the seeming life, power, substance and

84

---

intelligence of false conclusions. When the head is bruised, the whole is seen to be lifeless, powerless, unreal, unintelligent.

In the true immaculate conception and birth, the seed of the woman destroys all mental and bodily enmity. The Word is the child, so in the child born of the Holy Mary, everything   **Luke 1: 35** is transformed into Holy Spirit; and there is but one conception and birth for all children. In truth, all children are included in the immaculate conception, there is no other conception.

"Unto woman He said, 'I will greatly multiply thy sorrow and thy conception.' " The enmity between the woman and the serpent is the difference between the truth that I am now wise, as God's image and likeness, and the claim that I am to be made wise through experiencing opposites or contradictions within. Our study of Divine Science is vain effort unless we conceive and acknowledge the truth of at-one-ment for ourselves.

One who believes in physical and external causation has as many conceptions as he has opinions based in sensation and appearance.

"To bring forth children in sorrow," means bringing forth, as it were, in inequality and mortality. It is the belief of a loss of equality with God and immortality; and the conceptions of such are greatly multiplied about these results, and

85

---

about the body which they cannot see has any relationship to God.

"And thy desire shall be subject to the husband, and he shall rule over thee." Thy husband is (the Source), that from which thou art taken. Thou must stand side by side, at-one with Him, for thou are Being of His Being and existence of His existence. Man, male and female, are one in both Cause and Effect.

There is but one substance, hence but one Spirit and one body, and this truth shall rule mentally, for in Mary the true conception is unto salvation and immortality of the body.

It is Divine Adam created in the image of God, male and female, that has dominion; it is only the desires that are subject unto the husband. The image and likeness which is ever wedded to God shall rule over thee, instead of desire. And unto Adam He said, "Because thou hast hearkened unto the voice of thy wife, and hast eaten of the tree of which I commanded thee, saying 'thou shalt not eat of it,' cursed is the ground for thy sake."

It was Adam, male and female, that God commanded not to eat of the forbidden fruit, for He called their name Adam in the day in which He created them. "In sorrow shalt thou eat of it all the days of thy life."

For Adam, male and female, to rule over Eve,

the true mentality and power of generation, is to understand its true use, and comprehend its true meaning and relationship and presence as both Infinite Cause and effect. God saw everything good, hence to Him the ground was good; for the "sake" of the divinity of the whole, the ground was always good unto God. But Adam had hearkened unto the voice of his wife, which was, that good and evil were alike good for food, a thing desirable to make one wise. This was partaking of desire to be made wise by observation. Everyone who believes in good and evil, partakes of the same desire, and sees what he believes to be true in all things. Anything that is believed to have evil connected with it, is to that belief cursed, hence the Lord God said, "cursed is the ground for thy beliefs sake."

Thorns and thistles are the undesirable conditions sensed from the belief that there are two powers working in us, either for good or for evil. "Adam" stands for the whole race,   Mark 12 2 for mankind.. The works of the serpent and all the dark conditions which they include stand for all untrue mental conditions everywhere, at all times; conditions that contradict the real, true nature of God.

Jesus said, "Labor not for the meat that perisheth." Adam cannot wholly partake of that bread that cometh down from heaven while holding   John 6 27

the belief that labor alone entitles him to the bread of which he partakes. It is not right to believe that labor is the means by which to produce bread and that it makes and sustains life. Jesus says, "I Am the bread of life,' and to partake of this bread is to be the Christ. Adam, the world over and for all time shall labor to earn life, support, and to be made wise, until he returns unto the ground from whence he was taken.

Rom. 22:23   The ground from which Adam was and is taken, is the ground of Infinite Life, living substance and reality, the Source of all things. To return to Being is to know what true work means. Man is made, within this Holy ground, and God's free gift to him is Eternal Life. He was taken from the Ground by the false conception of creation and the contradictory belief that the things of God are both good and evil. But notwithstanding the false beliefs and judgments from observation, he shall return to this Ground, "for dust thou art, and unto dust thou shalt return." He shall come to know one substance, regardless of the conclusion of the senses. He returns when he refuses or denies a place to the false conception of duality and knows that he is the one true immaculate conception of God and "that that which is born of Spirit is Spirit."

Col. 3: 10   Thus he makes of, sees himself one new man and this slays all enmity within himself as it did in Jesus.

"Eve" means the idea of generation, to bring forth, to manifest Life. This is the true spiritual meaning, so she is spoken of as the mother of all living.

The Lord God making coats of skins and clothing them, shows that regardless of beliefs and opinions the Lord God protects, comforts and provides everything necessary. When we acknowledge the truth that we are the ground, being, atone with God, who manifests our living existence, we are ready to realize the perfect Edenic order and to dress and keep the garden in harmony according to the law of expression. "As a man soweth, the same shall he also reap."

The Lord God furnishes all food and raiment, but he who sows to the flesh believes in physical causation, and of that belief he reaps corruptible food and raiment. He believes he is clothed upon with mortality and that life is dependent upon death. This is the condition that exists when the Lord God says, "Man has now become as one of us, to know good and evil," immortality and mortality, life and death; "and now, lest he put forth his hand, and take also of the tree of life, and eat, and live forever: Therefore the Lord God sent him forth from the garden of Eden."

"Become as one of us," stands for the condition that is not as God is; that is, not as is the tree of Life. This duality is opposed to the unity of

89

Being and existence. It could not put forth its hand and partake of the tree of Life; it could not eat of it, and live forever. Therefore, the Lord God sent him, the seeming man that now says, "I Am, I Am, I Am," to all manner of beliefs and opinions and conditions, from Eden to till the ground, Being, from whence he, the true man, was taken.

Luke 11:17 — If this condition could be, remain in the Garden of Eden, it could be part of it and live forever, hence contradictions would be eternally in God and His manifestations. His kingdom would be divided against itself; a good tree could then bring forth both good and evil fruit.

Cor. 15: 54-58 — Adam had gone forth in his beliefs from the true being of what God is, and from true knowledge and wisdom, and from the truth that all is. good, to gain knowledge and wisdom through his labor of experiences, and thus to till the ground, instead of being it, and to gain eternal life through death, instead of being It, the free gift.

John 12: 23-26 — This seeming man of non-being, the one of becoming has no being in God, so he drove him out of the garden. But the man of God, male and female, is always and everywhere in the Garden of Eden. We are not to attempt to stand alone and be " as one of us," claiming to be something apart from God, for the creature cannot manifest life separated from the Creator any more than can

90

---

a branch bear fruit severed from its vine. God is the ground; He does not till the ground; the man of God is the ground, he does not till it; but serpentine opinions and conditions of desire till, cultivate, depend upon the outer, postpone their good and are ever seeking but never finding.

Eph. 6-17

"At the east of the Garden of Eden," means in the- light of pure intelligence and wisdom of Being. "There are cherubim and the flaming sword which turns every way to keep the way of the tree of life." "Cherubim" are blessings, and "flaming sword" is the word of truth that is with God and is God in the beginning. The blessings and power of God's possibilities keep the unity of the whole, the way of the tree of life, and no ignorance can ever enter into Being or existence.

Dominion is not a power to be attained through ages of evolution; man is created with dominion and is commanded to exercise it.

If thought claims two sources, spirit and flesh, it will believe in many causes, and will be divided in its conception between good and not good. The Bible is full of warning against this divided conception.

Only those among the Israelites who had wholly followed the Lord, entered the promised land; so it is only by undivided faith in God, which means in its infinite sense, to believe only in God and God manifest.

91

Lev. 19: 19

All the sorrow and suffering in the world is the belief in separation from God, "Apart from me ye can do nothing"; that is, belief in a power, a life, a substance and a mind besides God. Thought by this. belief sows the seed of all the misery in the world. Even in the old Jewish law, such false sowing was forbidden: " Thou shalt not sow thy field with mingled seed; neither shall a garment mingled of linen and wool come upon thee." To sow mingled seed, is to let thought hold to two opposite beliefs, good and evil; to wear the mingled garment, is to hold to two conditions in the body, strength and weakness, health and sickness.

Jesus declared, "If thine eye be single, thy whole body shall be full of light." If thy consciousness holds to one only, thy whole body will represent the nature of that one. "If thy soul be radiant, what can thy body do but shine." Such is the close connection between soul and body, for soul is the "eye" that first sees the purity and perfection within Spirit, and recognizes that the same perfection must belong to body, which is born of Spirit.

To look only at Spirit and to know that Spirit is all Truth, hence that the true good of all things, the true Life and Substance of all things, is Spirit and that like produces like, is to have the eye single, is to have the soul radiant and to see the whole

92

body full of all good. Looking to the external as a cause is the beginning of thoughts sowing mingled seed. The Tree of Life represents Man as he is seen and known by thought that recognizes and faithfully holds to Spirit as all Source and Cause. Soul and body are then seen as the perfect image and likeness of Spirit, and the oneness of Creator and Creation is realized.

The Tree of Knowledge of Good and Evil represents the same man as he appears to our thought when it forgets the One and Only Source and claims form to be a source when it looks upon the visible as the maker of our happiness and wisdom. Divine order, or law, is thus reversed it moves always from within outward --- from Mind through thought to Word from Spirit through individual soul to body from the vine through the branches to the fruit --- and never otherwise.

To turn to "Egypt" for help, is to partake here and now of the Tree of Knowledge of good and evil, for it is to look to form, to things we can see and handle, for the support of our life and health. This is a mistake, a missing of the mark, a sin, for only in Invisible Spirit Omnipresent, Omnipotent, can we find any Cause for good.

Adam and Eve partake of this false belief because thought is in its growing condition.

Israel turns to this false belief of good for the same reason.

93

Eden and Canaan are the perfect consciousness towards which this immature thought is pushing. In the Garden of Eden, or the perfect consciousness in which thought is placed, Truth says, "Turn only to the Spiritual idea of life; do not look upon form as a source; all your good, all your wisdom must be found in Spirit. In the day, light of understanding, that you turn to another source but Spirit, and receive your conception of life from the appearance, you shall surely die." Such turning from Spirit obscures our vision for the time of that understanding and banishes thought from consciousness of its place in Spirit, from a knowledge of its relation to Spirit; "Therefore, the Lord God sent him forth from the Garden of Eden to till the ground," or to work out his own belief in material life and substance. Man has chosen to judge from appearances, to turn to the outer for his supply; now he must go on in that belief until his own experience, "the end thereof," proves to him its fallacy.

Matt. 10: 34

A flaming sword keeps the way of the Tree of Life. When man partakes of this tree, he can reach it only by passing through the "sword."

The flaming sword is a symbol having the same significance as the "consuming fire."

Each soul must pass through these to reach consciousness of Eternal Life in Spirit; every false conception we have held of life as being anything below the

94

Infinite must be cut off or burned away from that thought which would lay hold upon the eternal and unchanging idea of life.

The serpent of desire bids us seek our wisdom and good in the external in opinions and beliefs of men. We first look upon, then long for, then embrace, and so "comes sin into the world and all our woe." Not by the mistake of one, but by the mistake of our own immature thought, turning daily to the external as a source.

Gen. 2: 16-17

The tree of the knowledge of good and evil was not in the garden of Eden. The "Tree" of which our thought is directed to take is the true idea of life as judged from the standpoint of Spirit. The "Tree" of which thought is forbidden to take, is the opposite conception of life based upon outer things.

God made both Canaan and Egypt. He made all things Invisible and Visible. The command is that thought shall always look to and be guided by the Invisible, to know Mind as the only Power, Life or Substance. The visible will come forth then in beauty and perfection but the visible is not to usurp God's place, the first place in your thoughts, it is not to be put as cause, which Spirit alone is.

The whole visible universe is good; it rests in the very bosom of Infinite Love, for this tree representing visible life is "planted in the midst

4

95

of the garden"; it becomes the tree of the knowledge of good and evil only when thought partakes of the idea that this visible world is the origin and source of supply for his body; believes in two sources, spirit and flesh.

While man's thought is turned to the outward, acknowledging it as a cause, failures come; the cry of lack and poverty goes up to heaven. Why? Because man is looking to that for supply which is not a source.

Held as a work of God, or result of good, the visible is seen as the manifestation of God, perfect and good. Entered into as a Cause of good, the only Cause of all good is forsaken and soon a sense of lack and want is felt, which will drive us at last to seek for that which is an unfailing source of good, even the Infinite Unlimited Presence of Divine Mind.

The Garden of Eden represents the possibility in man of being God-like. The serpent represents desire in man's mentality, desire has place in undeveloped thought and impels man to seek. The Garden of Eden is verified in the land of Canaan; Adam and Eve, in Israel; the serpent, in the famine; the flaming sword, in the plagues; cast out of the Garden, in the going down into Egypt; tilling the ground, in the bondage in Egypt.

96

Return to the Top Return to the Top

THE RIVER.

I. The river in Eden has its rise in Infinite Being, the ground of Infinite Substance in the possibilities of God, it flows out into all expression of God, in the hosts of heaven and earth.   Ps. 46: 3

"From thence it was parted, and became into four heads." It watered the whole of existence, it is the power of God to live all things and demonstrate all power at the same time.

The name of the first river is Pison. First, Ehhadh, signifies joined together, many united into one. As the different members are united into one body, so this river represents the truth that the living body is united with the one Spirit. River signifies action. Pison denotes to overflow creative action and circulation. It compasseth the whole land of Havilah. It flows through the whole of the creative action, where there is gold.

Havilah is to create, re-form, to supply strength Pison covers all these. It is the power of God to create continually. The gold, bdellium and onyx stone stand for the richness and fulness of power of this river in creative action and circulation, and the abundant unfailing and eternal supply of the source from which it flows.

It may be illustrated by the infinite variety of thoughts circulating in Mind, and actualized in the body in the circulation as performed by the heart and its tributaries, thus demonstrating life,

97

vigor, and all good and maintaining mental and bodily health, harmonious action and beauty of form.

The name of the second river is Gihon Second, Shani, means to repeat, to do over again. It is twofold and in God's relation to finished creation it stands for the fact of the individual mentality's receiving from God and giving the same expression, its conceiving the truth of God and giving birth to the same. Gihon is that which bursts forth into constant activity, hence this twofold river has been called the "double tongue". The individual receiving from the Omnipresent Spirit and giving forth the same is symbolized in respiration the lungs, twofold in structure, continually repeating the process of inhalation and exhalation. So it stands for God breathing the breath of life into man, not from under heaven, but from within Himself.

The land of Ethiopia is the home of the most primitive tribes, the river compasseth the whole land, the spiritual-meaning of which is the ability of the Lord God to breathe the breath of the spirit of life within and to be the life of everything, including the simplest most rudimentary forms of life. Primitive pertains to the beginning or origin, or to early times --- original, first, as the primitive state of Adam.

The name of the third river is Hiddekel. Third,

98

Shelisha, is the rule; the rule river directs, determines, chooses. Mentally, it keeps the way of the tree of life, and judges righteous judgment.

It is actualized in the body in digestion which presides over, decides and approves of the food partaken.

What is spiritually true is actualized and is literally true, for the outer is as the inner. East is light, priority, and Assyria is success. Thus, these words show the importance of right judgment according to the unity of the whole, that we are to see God manifest in action of the body.

The fourth river is Euphrates. Fourth, Rebhii, is an emblem of creative power. Euphrates signifies to enlarge, to weave together, commingle. It is the word of God, the voice of heaven. The perpetual covenant that the one living and true God keeps with all creation for perpetual generation.

The one universal holy, immaculate conception, mentally received and bodily actualized in marriage, which is the oneness of spirit and body, and the oneness in substance of bodies. Heb. 8:10-11
It says, "Let us make man," so time man that the Lord God forms of Infinite Substance has all this truth actualized within Him. John 2:1

The river is flowing out of Eden and watering the garden throughout, in every part at this time. If the eye is single to truth, the whole body is full of light, eternal day, God speaking. Matt. 6-22

99

Return to the Top  Return to the Top

CHAPTER 2. - HISTORY OF THE ISRAELITES TRACED FROM THREE EPOCHS: CREATION, FLOOD, CALL OF ABRAHAM ---

A. The Bible from Adam to Jesus is an account of man's spiritual unfoldment.. - p. 100

History of Israelites Traced from Three Epochs; Creation, Flood, Call of Abraham

A. We hope to get the one thing we wish to have in thought as we read the Bible. We have seen that the first Chapter of Genesis is a revelation. If you were in the dark and a light broke and kept growing more and more clear, you would see things you had not seen before, and you would think they were new creations, but it would be only that the light was bringing into your vision that which was there.

Creation took place in the beginning, and the revelation is simply God's being revealed. There is nothing else to reveal, and while God is being revealed, something is taking place and that is that the Self is being revealed. Everything is being revealed in the light --- one sees nothing in the darkness. The result of light is expansion of every energy, of all that is in you, of your consciousness; and the result is manifestation, a thing is manifested. Word and thing are just the same. The word of God is a thing, and a thing is the word of God, the word made flesh.

As we go along, we have more light, therefore, more expansion, and finally the great vision is

100

---

man in God's image and likeness; this is revelation. The Universal reveals itself, the individual receives the revelation. The Truth is ever lifted up and the conception about it passes away.

Whether the stories of the Bible are literally true or not does not make any difference. The chief reason that this story, termed the "fall of man," is so well known so far as the   Gen. 2: 3 the scriptures are concerned, is because of certain expressions of Paul. The prophets do not say anything about it, nor is it referred to in any other place in the Old Testament.   Rom. 5:12 19 Jesus does not mention it. Adam and Eve represent the race to us in its first spiritual consciousness, interpreting from its immature conceptions and seeking wisdom and all   1 Cor. 15:21,22 goodness just as fast as they came to see any value in goodness. We do not think they were the first human beings. We think they represent man's first glimpse of spiritual consciousness. (We are told that Adam was a living soul.) We can conceive of people existing before them, but without any spiritual consciousness. We think of them --- Adam and Eve --- as the awakening soul, the first desire for wisdom and goodness, in that consciousness it was very easy for them to turn to the external as the source of wisdom.

The first Adam, a living soul, a little conscious of soul; the second Adam, a quickening spirit, the spirit that giveth life. The difference between the   1 Cor. 15:45-47

102

---

first and last is simply a difference in consciousness.

The Bible, from Adam to Jesus, is an account of man's spiritual growth. In immaturity, mistakes are made, but the mistakes of a child merit no condemnation, only training is needed. Sin is simply a missing of the mark.

Matt. 6: 23

John 3: 19

Now, this breaking of light into each individual mentality is the first step in development, unfoldment. This light is calling the light out of the darkness of our ignorance. No man to till the ground but the one who partakes of the forbidden fruit; when our thought is mixed we partake of the tree of knowledge of good and evil. In the midst of the Paradise of God.

Gen. 4

The fourth chapter of Genesis represents the dual belief of good and evil, that God and man are separated and unlike each other in their nature. Adam and Eve, who stand for the whole race, have a desire to be made wise through contradictory experiences, and they begin their effort to acquire wisdom, all that they are to have or possess; they make the mistake of trying to obtain instead of attaining, of acquiring instead of Being.

Every conception brings forth its offspring. Like begets like. The flaming sword will never let the ignorant conceptions of the past go into Eternal Life. The Adam and Eve consciousness

102

brought forth Cain and Abel (duality). All start as Adam and Eve, and Cain and Abel are the offspring of that consciousness.

Abel, the feeder of the sheep, the good shepherd; Cain, a tiller of the ground, just what the man was outside the Garden of Eden. "And in process of time it came to pass, that Cain brought of the fruit of the ground an offering unto the Lord, and Abel also brought firstling of the flock." The Lord God had respect to Abel but not to Cain.

Gen 4: 2

A tiller of the ground would not be typical of as spiritual a consciousness as a keeper of the flocks; Abel, spiritual nature within; Cain, that which turns to the outer.

Eden, our consciousness. A dual thought takes us out of Eden.

The sin, falling short, of Cain, which "lies at the door," is that he is always acquiring something from an outside source which is his offering unto God, but he never offers himself.

"And it came to pass when Cain and Abel were in the field, that Cain rose up against Abel and slew him." This has been interpreted as two natures within us, material and spiritual, but this belief is only possible in the process of our development.

"Brother's keeper" means Cain's belief in separation. "Cursed is the ground"; that which

4a

103

is cursed is rejected, it cannot be at-one with that from which it is cursed.

The mark that the Lord sets upon Cain lest any finding him should slay him, is a heavenly sign of protection. The mark is understanding or knowledge of Truth. He, man, represents all possibilities of growth in knowledge and understanding of the Omnipresence of God and himself as the expression of all that God is; yet he dwelt in the land of Nod, wandering land or flight.

The story describes race unfoldment in this killing man began to see for the first time the sacredness of life, the right of life.

After beginning to glimpse the value of a soul, killing became wrong, this was a sign of development. It was the awakening of the moral sense. In the first day is living soul infinitely provided for, but man failed to see the light of the Truth.

Gen. 5: 1-3

From Adam to Noah are ten generations. The fifth chapter is the book of the generations of Adam in the day (the light of understanding) in which God created man.

This represents the generical and real state of the first day of creation.

The condition of Cain and Abel represents the earliest product, or first stage of unfoldment unto every one who believes in good and evil, This record is for all time.

104

THE FLOOD.

Did God see that man was evil? No! Man never has been evil. God saw that the imaginations of his heart were evil. The imagination of man's heart is evil from his youth because of his youth.

The flood is coming to each one of us every moment. It is the flood of light and Truth. It is going to wash away the darkness of my ignorance, it is going to show me myself in God's image and likeness. Gen. 6: 5-7

Gen. 7:21, 22

"All flesh perish from the earth," the thought of the un-Godly will perish in the flesh. Everything that moves upon the earth shall be destroyed, everything that is low in its tendency.

The "dry and thirsty land"; the place where God is not recognized. Every thought which was resting in that place where God is not recognized was to be washed away in the flood.

Ark --- consciousness of our true Being, safety. Windows of heaven --- eyes of the soul opened. Waters --- possibilities.

Flood --- fuller realization, washed away every conception, purified thought. Noah waited forty days. Birds sent out --- perfect thoughts; mountains --- high visions; rainbow of promise represents law and order in nature, a symbol of God's promise. Noah stayed in the ark until God commanded

105

---

him to go forth and let his light shine; his first act was to build an altar to the Lord and offer sacrifice, then he received God's promise and God's "promise" is law.

The "Tower of Babel" means man trying to work his own way to heaven, trying to get through his own personal effort that which God had already given him.

John 16:12-13; 17:3   Personality is our greatest enemy. Self, individuality, is divine, but the personality which we have let cluster around the real self is the enemy. This personal belief always tries to build its own way to heaven. Our effort, our motive, is to be impersonal. The Tower of Babel we try to build is sure to end in confusion

Return to the Top   Return to the Top

B. Abraham.. - p. 106

B. Genesis is like a cord twisted together, the interwoven strands appearing and disappearing. It is divided into two parts, chapters 1 to 11, the real Genesis lessons and chapters 12 to 50, the history of Israel 's ancestors, the patriarchs.

We find in the stories of Abraham an illustration of different sources of information. In Genesis, 16th chapter, we read of Hagar as a mere slave with no rights, driven out by. Sarai before Ishmael is born. Abram shows no feeling with regard to her; she was the property of Sarai, who could claim her children as her own. In Genesis,

17th chapter, there is no reference to Hagar or Ishmael's being cast out, but Ishmael is mentioned as Abraham's son of thirteen years when he was circumcised. In the 21st chapter, Ishmael is born; it is not until Isaac is weaned that Sarah wished to have- Hagth'i cast out, and it was very grievous in Abraham's sight. He rose up early and gave Hagar food and drink. Paul, in Galatians 4:24, calls the story of Abraham an allegory! "Which things are an allegory."

We also find in 12th, 20th and 26th chapters of Genesis the same stories applied to different people. Abraham was called to serve God, the within; "to leave his country,"   John 8:35 personal and race beliefs. Kindred, personal beliefs; Father's house, inherited beliefs. Abraham obeyed the call, promise followed, more law than promise. Cause and effect. We are to cast out everything of the personal belief, hold only spiritual birth or nature, nothing but divinity; we are divinely human.

Right after Abraham had cast out Ishmael, he was called. The offering of Isaac means that we are to give up that which we love as well as that which we do not love.

The meaning of "sacrifice" is to make sacred, make everything sacred, the body sacred to life. The promise was to Abraham and his seed. "And I will establish my covenant between   Gen. 17: 7 me and thee

and thy seed after thee in their generations for an everlasting covenant, to be a God unto thee, and to thy seed after thee." The seed of Abraham was for a long time limited to the Jewish people, the descendants of Abraham, but the New Testament wipes out that thought entirely.

Rom. 9: 6- 8	Gal. 5: 6
Rom. 2:28 -29	Gal. 6:15
Rom. 3:29 -30	John 8:39
Rom. 4:12 -13	

Seed, possibility that is in the individual.

Ps. 126 :6	Gen. 1:11
Isa. 66:22	Eccl. 11: 6
Rom. 4:16	Gen. 17: 7

All the above are references to the seed of Abraham. They take you out of the letter, out of the flesh. Also Hebrews, 11th chapter, tells what is done by faith.

Gen. 15 8

Abraham was faithful, but be asked for a sign; he doubted. A cloud was cast on him. Doubt leads us on, and if we yield, we suffer. Confidence and trust must accompany faith, work, action and obedience.

Gen. 20:17

Gives the first recorded instance of divine healing.

Gen. 32: 24, 32

The story of Jacob's ladder. The people believed in a personal God and the ladder was bringing heaven and God near to man, unity, a comparatively

high step for Jacob. When he waked, he saw God in that place. "Jacob carried with him the mark of the struggle throughout his earthly career. The angel touched the hollow of his thigh; he was lame. He could never walk again in earthly things with the same step as before. He had seen the Truth, and could never again unsee it; his whole viewpoint had been altered and no longer understood by those around him; to those who have not wrestled with the angel, he appeared to walk lamely."

New Name, name as result of his wrestling with angel, his own soul, the work of each individual. We wrestle until the day (light of understanding) begins to break, until the I Am   Rev. 2: 17 is born; born of the Water and the Spirit.

Return to the Top   Return to the Top

C. Israel turning from Canaan to Egypt.. - p. 109

Gen. 41:56-57 and 46:31

C. Gen. 42:1-2; Gen. 43:1-2

Gen. 45:9-13, 17-18

Gen. 47:5-6-7

A famine in the land of Canaan induces Israel to send to Egypt for supply. This corn is soon exhausted, and again, a second time, Israel seeks help from Egypt. From this second turning to Egypt for help, an invitation comes from the ruler

109

of Egypt to Israel that he come and dwell there; the promise is made, "Ye shall eat of the fat of the land." This call from Egypt is heeded, and the third step taken. Israel moves with all his family to the new land where plenty is and feels well satisfied with the change, for we find him a little later blessing Pharaoh. Let us find an interpretation of this in our own spiritual experience.

A "famine" is a dearth, a barrenness, a lack. A famine in Canaan would signify a serility in Consciousness of Spirit, a lack of confidence or faith in Spirit. It comes in our experience when our thought does not see anything substantial in Spirit, does not understand how the Invisible can be a source of supply for our needs.

There is never any lack in Spirit; the lack is in our realization of Spirit's Power and Presence, our blindness to its Fulness which is for us.

Phil. 4: 19   It is written, "But my God will supply all your need," but thought in its ignorance does not hear aright this promise of supply from within, hence sees no possibility of its fulfillment here and now.

Gen. 3: 19

Man does not believe that Spirit can supply the earth with every good thing; he sees nothing in. Spirit to satisfy his present desires; he cannot receive the unlimited assurances of Truth which comes from within, for he does not hear them; so

110

---

he begins to search in the outer, or material, for his daily supply. Thus is man under that law of bondage which saith, "In the sweat of thy face shalt thou eat bread." He knows not how to claim for himself from Spirit, "Thou givest us our daily bread."

Can the Spirit, indeed, give us our bread and our meat, our clothes and our homes? "Consider the lilies how they grow," said one who knew! "They toil not, neither do they spill, yet I say unto you that Solomon in all his glory was not arrayed like one of these." — Matt. 6: 28

We do not understand Spirit's word; this is the cause of the "famine." "It shall come to pass in that day, saith the Lord, that I will send a famine in the land. Not a famine of bread nor a thirst for water, but of hearing the words of the Lord." Spirit is fulness omnipresent. "All things are yours." The fulness of Spirit, all that Infinite Spirit is, is yours; but you must claim your own. — Amos 8: 11

Claiming good does not bring good to us or make it ours. It is ours before we claim it! Claiming the Truth only makes us realize our eternal good. If our eyes are blinded we shall not see, though we stand in the midst of light; so, too, if the inner eye or thought is blinded or ignorant, it will lack all things, though in the midst of plenty.

111

Thus it is written, "It shall come to pass in that day, saith the Lord God, that I will cause the sun to go down at noon, and I will darken the earth in the clear day." Accept the thought that man is spiritual and not material, that he is centered in God, is in the very midst of Fulness and has a right to all good, he has been given perfect freedom and infinite supply; he is granted without asking, perfect strength; all things are his if he will only accept all things. — Amos 8: 9

Suppose, then, you see this one so fully endowed acting like a slave instead of an heir, imagine him groaning with pain, despondent from a sense of lack and need, burdened with care and anxiety, suffering from poverty and fear, lacking all good, would you not exclaim, with the Master, "How is it that ye do not understand? Having eyes, ye see not; and having ears, ye hear not?" When thought, in its blindness, sees no help in Spirit, cannot see the Source within, where shall it turn for its supply but to the external? Thus

Israel: today, not sinnners, is doing this! "My people have forgotten me days without number." — Jer. 2: 32

Christians have turned to Egypt for help because they know not how to take possession of the riches of Canaan, their own land. "Who is blind but thy servant, or deaf as my messenger that I sent? Who is blind as he that is perfect and blind as the Lord's servant?" This blindness is lack of — Isa. 42: 19

112

---

understanding which moves thought to look to the visible or form as a source, instead of to the Invisible.

It is hard for us to understand that the Inner, or Invisible, is All Cause; that from it alone each individual must receive all things, hence that we cannot, in Truth, find any cause in the external.

When we ignorantly claim that the visible is a Cause, and seek to gain our good from thence, our Israel, or Spiritual thought has turned to Egypt.

Today we find ourselves in Egypt, dependent upon material things, guided by human opinions of good and ill. How did we come there? Just in the same way that Israel went there. First, Israel sent to Egypt and bought corn. It is reasonable to think that in this first turning to Egypt for help no idea of ever going into Egypt to dwell was entertained for a moment. Israel brought the corn out of Egypt and fed on it in. Canaan. Have we not been trying to dwell in Spirit and yet be helped and sustained by what material sense calls good? Have we not our Sundays and prayer meetings in spiritual consciousness and our everyday life, perhaps, unconsciously given to material sense?

Can we live both in Canaan and Egypt? This is the first step that will lead us into bondage in Egypt. Jacob was doing this only as a temporary relief from the famine. So on earth our spiritual

113

thought has felt, "I'm but a pilgrim here, heaven is my home." And while I dwell on ,earth, all spiritual power for good cannot be given; I must, for the present, he sustained from the earth, then, when I leave earth for heaven, Spirit will be my full supply. So the "first man is of the earth earthy," for this is the first, or infant belief that man has of himself. But his knowledge shall increase.

John 4: 14; 4: 32   We find this first help from Egypt soon exhausted. As all help from without is limited, it soon fails. But the master of life says, "Whosoever drinketh of the water that I shall give him, shall never thirst again." He understood the full supply of Spirit, and could say, "I have pleat to eat that ye know not of." No more poverty, no more ,starvation and distress when this truth is realized; the true supply is for all alike.

When the first supply from Egypt is consumed, what can Israel do but send again? So, the second step in the descent into Egypt is taken, making the third or final step easy. Egypt is friendly, and its invitation, "come unto me," with the assurance of its ruling beliefs and opinions, "I will give you the good of the land," revived the Spirit of Jacob, we read, and he went down into Egypt Nvitli his family to dwell. By degrees, do we turn from Spirit to enter at last wholly into materiality, or material sense. It was thus with

114

our mother Eve; she first merely listened, gave heed to the voice of desire, the tempter; next, she is led to look upon the "forbidden fruit" with desire in her heart; then she sees its beauty, believes in its power to give satisfaction and pleasure; third, and last, she takes and eats.

Where and when did this occur? Here and now within us! Not conscious of our fulness, we desire, then we look at the beauty of God manifested in the visible, we mistake it for the Source of goodness when it is only a result. Whence has that beauty and goodness originated? For to that same source must we go, else the supply will soon fail. The external is not a source of anything; it is a result, an effect of an inner Cause. The inner is inexhaustible; it is the source of supply to every external thing. To that Limitless Source we must turn wholly and see its fulness omnipresent, if we wish to feel no more the "famine in Canaan." It fills every longing of soul and body with satisfaction.

We cannot receive any permanent good from without. We cannot receive any permanent harm from without. The outer can neither help nor harm us. "All my Springs are in Thee."   Ps. 8 7: 7 Mine eyes hast thou opened. I acknowledge Thee in all my ways; Thee only, Source and Cause of all that exists,

115

Israel is thought that trusts in Spirit. Egypt is thought that trusts in materiality as a reality.

D. How we may know the Divine Will.. - p. 116

1 Sam. 8:5-10, 19-22

D . 1 Sam. 9:16-17

1 Sam. 10:1

Prov. 14:12

"There is a way which seemeth right unto a man, but the end thereof are the ways of death." How, then, shall we know the right way if that which seemeth right, is not so? We may know by the end or result thereof. "By their fruits ye shall know them." "Like begets like" is the law and that conception which is heaven born will bear the impress of heaven in its results.

Gen 46: 2-4

The move into Egypt seemed to Jacob to receive its Divine sanction. We read, "And God spake unto Israel in the visions of the night, and said, 'Fear not to go down into Egypt; I will go down with thee, and I will also surely bring thee up again.' " Do not these words show Divine endorsement of this move upon the part of Israel? Does it not seem that it is the will of Spirit that we turn to find in the external a supply that it cannot give? We are forced to say, no, indeed! Spirit, by its omnipresence, is ever giving us full supply, but we, not seeing it, turn to external for help we have sought in Spirit. Then, what significance have these words to Jacob? Let us carefully

116

consider the saying, and in today's understanding search for its meaning. These words came to Jacob "in the visions of the night." This is when personal thought hears, in the darkness of its own ignorance. It does not indicate clear hearing; it certainly does not suggest that the Divine Will was perfectly understood, but rather, was received through a dull consciousness.

A subtle temptation is exposed to us in this. Do we not often hear Divine Will as it is colored by our own desire? Surely, if we go to Divine Wisdom and claim our intelligence wholly in it, without any holding to individual will or desire, such pure thought will receive perfect consciousness. This "hearing" will not be in the "visions of the night," but in clear day. This, then, reveals that Israel heard not Divine Will clearly, and understood not what he heard.

There are some who say, "Does not God give us all these things for our present help and supply? I pray for God's blessing on them, and then feel that I do his will in using them for my good. I find my good health in climate, food, exercise and herbs, all of which God has surely given us, and by His will we use them." Certainly this is the way that seemeth right; but what is the end thereof? Half of the world is starving and freezing; all the world is suffering and dying, "the end thereof is death." The fruit or result of such belief

117

proves it not to be of Divine Will; man is interpreting Divine Will by the light of his own desire. "Be not deceived, God is not mocked. Whatsoever a man soweth that shall he also reap." This is the Truth contained in the belief of eternal punishment; for it is eternally true that sin, or ignorance, brings upon us its own results, and will until all ignorance is destroyed. Later in our study, we find what was the end of this move into Egypt, and what is the result to us of the turning of our thought into the visible as a source.

Again, in the words heard by Israel in his "night visions," was this assurance --- "I will go with thee into Egypt and I will bring thee up again." What do we learn from these words if they do not show Divine pleasure in the going to Egypt? Simply a truth, namely, that God is Omnipresent, always with us, even if we turn from Him! There is no move we can make unaccompanied by Divine Presence. "If I make my bed in hell, thou art there," does not teach us that the Divine delights to have us go into torment. So these words heard by Israel are not an encouragement for him to go down into Egypt, but an assurance that even though he choose to go away from "Canaan," Divine Love is still with him. What for? "I will bring thee up again."

Though one sink to the depths of ignorance, Divine Power never forsakes nor turns away

118

---

from him, but waits to bring him up again. " Thou wilt not leave my soul in hell." Hell is the purifying process in which God as "a consuming fire" is purging each soul of its false conceptions. Truth knows that no one will stay eternally in such depths of ignorance, for God is omnipotent. Sooner or later every one will turn again to Spirit and find therein its fulness and freedom. "Every knee shall bow, and every tongue confess."

When thought turns from Spirit as the only Source and goes into what personal sense offers as help, Israel leaves Canaan for Egypt. But Spirit is ever with it, even while it turns   Isa. 45: 23 away, Spirit's Presence is the drawing power that will bring it home again. The home of Spiritual thought (Israel) is consciousness of Spirit (Canaan). In other words, from the Bible we are forcibly told that Spirit does not in any sense endorse the turning of Israel from Canaan to Egypt, nor promise any blessing upon such a departure.

We read, "Woe to them that go down to Egypt for help, that stay on horses, and trust in chariots because they are many, (we are tempted to turn to material sense because the   Isa. 31: 1, 3 greatest number seem to be on that side) and in horsemen because they are strong." (Strength seems to be with the visible rather than with the invisible. Many of   Isa. 31: 3

119

---

us are tempted by this.) But they look not to the Holy One of Israel (the Divine Nature of thought), neither seek the Lord. Now the Egyptians are men and not God (are human opinions and not true knowledge), and their horses flesh and not Spirit. When the Lord shall stretch out his hand (when the Divine Truth and Power shall be recognized), they all shall fall together. In this, the third verse, interprets the first, making it read spiritually thus: Woe shall come to them that go to men's opinions for help and not to Divine knowledge; who believe in flesh instead of Spirit, and trust in the visible because it seems stronger and better able to satisfy than the invisible; who look to the outer as a source of help and strength rather than to the inner.

Again we read, "Woe to the rebellious people, saith the Lord, who take counsel, but not of me; who cover with a covering, but not of my Spirit; (who seek wisdom from appearances, trust in the outer and put man's opinions above the truth of spirit) that walk to go down to Egypt, and have not asked at my mouth to strengthen themselves in the strength of Pharaoh and to trust in the shadow of Egypt; (who make great effort to conform to human laws, not recognizing Divine Law as all, who claim their strength to be according to ruling opinions in material sense, and put their trust in that which is but a shadow of good);

120

---

therefore, shall the strength of Pharaoh be your shame, and the trust in the shadow of Egypt your confusion."

"Lo, thou trusteth in the staff of this broken reed, on Egypt, whereon if a man lean it will go into his hand and pierce it. So is Pharaoh, King of Egypt, to all that trust in him." How plain these words are. Spirit is the only Power that can help man; Spirit is the only Source that can supply man; Spirit is the only Cause of Life, Strength, Health and all good for man; therefore, satisfaction can never be found in the external until thought seeks only from the inner and by the law Divine expresses in the external what it has received from within.

One stands by the river of Life, on each side of which grows the tree of Life "whose leaves are for the healing of the Nations." Within my own being is all the sweetness and perfection of Eternal Life, which finds an image of itself in life's stream by which I pass. Mistaking the beautiful likeness in the stream for the original, I seek to obtain more perfect conditions of life from this shadow; and in doing so, lose sight of the true substance of my being; also soon lose the shadow.

From this, I learn that eternal and changeless Life, with all its perfect and peaceful conditions, with its fulness of substance all satisfying, must

121

---

be received from within and be held firmly in my consciousness, if I would see its perfection and peace fully and beautifully imaged in daily life. If I try to gain my peace from the image and likeness, from the visible which is God's manifestation of Himself, I leave the true and original Source of good, and soon find lack everywhere. The visible is the result of good, but not the cause.

Return to the Top   Return to the Top

E. Israel's Bondage in Egypt.. - p. 122

E. Canaan is consciousness that trusts in Spirit.

Egypt is thought that trusts in materiality.

After our thought has once been turned to the outer, and we are led to gain our supply through personal effort, "tilling the ground"; thought is more and more weaned from dependence upon Spirit until it finally accepts the opinion and beliefs of material sense. Thus we mingle with people _of that land as is said of Israel, and likewise bless the material means we use.

Jer. 2: 13   "My people have committed two evils; (made two mistakes) they have forsaken me, the fountain of living water, and have hewed them out, cisterns, broken cisterns that can hold no water." They have left the true Source, and are believing that to be a source which can hold no good for them, these are their two mistakes.

One may ask, "Has the individual nothing to

122

do; no effort to make"? Yes, he has much responsibility; not to make results, however, but to receive the idea from the Universal Divine Mind, of that which is eternally made. Divine Truth and Law must govern his actions. Look where we will in nature, we are impressed with the Divine Law and Order that pervades all things. Law is universal; the individual must understand the Law that governs his thought, or else thought Will move ungoverned by Truth, and perfect results cannot follow.

Thought has a place in the Divine plan, has a work to do. Its place is between the Great Invisible and the Visible, the Infinite Mind and its word. Thought has been compared to a branch of a tree, and words or bodies, (the visible forms) to the fruit. The place of the branch is between the parent tree (source of all fruit) and the fruit. The branch does not make the fruit, it simply bears out from the tree that which is necessary to make fruit. The branch must always maintain its connection with the tree, and must draw entirely from the tree, if it would bear good fruit; for all the good of both branch and fruit come from within the tree. If in any way the branch separates from the tree, both branch and fruit die. By this illustration we may learn how Invisible good is made Visible by thought abiding in and drawing from its Source, which is consciousness

123

of Spirit. The responsibility of the individual is to see, to realize what God is and does. Imagine a tree with strong branches and rich looking fruit. Suppose you hear the branch saying, "I can support myself; here is this beautiful fruit to satisfy my desires (looking to Egypt); I shall give up my attachment to the tree and feed upon this good fruit." See it sever its connection with the tree and begin to live upon its fruit (going down into Egypt). For a time it seems to flourish, but its doom is inevitable; before long both branch and fruit wither and die, Israel is in bondage. There is no mystery at all about the ills of life. They are the withered branch and its dying fruit. They are the result of thought claiming a life apart from Spirit.

The "prodigal" represents that condition of thought which leaves its home in Divine Consciousness, separates from its source and tries to find in the outer, which is called the "far country." its strength and substance. We know how at last, starving, it returns to its home.

Material sense, which is ignorance of Divine Law, claims that the visible or material is good to make one wise, to make one strong, to make one happy, to make one well. But, the same sense declares that the material also can make one weak, unhappy, and make one sick! Such belief, if accepted by our thought, makes the external a

124

source of the knowledge of good and evil, and this is the forbidden tree. We know the true Source sends forth only good, for it sends only what It itself is. The same fountain cannot bring forth sweet and bitter.

Strip the allegory of Eden of its disguise and read its truth in the new tongue. Spirit is now revealing the same truth to us, saying, Man is a spiritual being; his consciousness of Me begins as a seed in the soul, capable of all Divine possibilities, but undeveloped (Garden of Eden). I am moving him out of darkness into light, out of ignorance into knowledge, gradually, as lie can receive it. Let there be light in his consciousness that the Truth of himself may be revealed to him; let this man look always to Spirit, since he is My image; he shall come to know himself by knowing 111e. Let his soul witness what I am, then shall it receive all light and understanding from Me, and see all things as My likeness.

This seed of knowledge is implanted in his spiritual consciousness (Garden of Eden), which is nourished and kept by My power. I give this seed, which is My Divine Nature, into his keeping; I put the source of all goodness (Tree of Life) within him, and from this Source springs perfect fruit, but let him look always to the Source; to the within (or consciousness) for his peace and wisdom. Let him not turn to the external and

125

dwell there; let him not be allured by the beautiful form of Life, but cling ever to the Source, if he would be continually satisfied.

In the Spirit is all Cause of Perfect Life, Substance and Intelligence --- all source of strength, joy and health; and not from the external form can these be received. Soul is to receive from Spirit and give to body. I Am the only supply in the universe. From the I Am, the whole (body) or Visible universe must draw its substance and life. Let Men obey my words, understand my law and follow it, drawing everything from Me, Spirit, and not from "flesh." Then shall perfect harmony be found in body, for every thought and word will truly represent my Divine Nature. If a man disobeys, or disregards this Truth and makes the "body" a source and cause, such ignorance separates him for the time from the consciousness of Spirit. The body becomes his master, he must serve the external to whom he has made himself "a servant to obey"; for by calling the external Cause, he has put himself under external things.

Hence, although I have "put all things under his fleet," he subjects himself to all things, and serves in fear and trembling, seeing a law of good and evil, of life and death, of rest and unrest. He must work out his own salvation until this mixed belief comes to an end and his thought, turning back to the true and only Source, sees one nature as all;

126

---

sees the real Cause to be good, life and peace, for it is God. This seems to be the truth as revealed by today's light in the figurative language of Creation.

Jesus gives the same teaching in his parable of the wheat and tares. The Master planted only good seed. The Garden of Eden is the field, Man's soul that has within it the possibility of producing or expressing the perfection of Divinity. "While men slept" an enemy came; this was a dream caused by ignorance, or immature thought; the enemy appears only while we are asleep to truth. Matt. 13: 27

When we awake, we question, "Whence has this field tares when thou hast sown only good seed"? Jeremiah hears the same question, "I had planted thee a noble vine, wholly a right seed; how then, art thou turned into the degenerate plant of a strange vine unto me"? Jer. 2: 21

As light increases, true judgment comes, and separation then begins to be made. Knowledge of Truth separates within us the wheat from the tares, garners the wheat planted in the beginning (all good in all is eternal), burns the tares, for the light of God revealing Infinite Good as all Truth, consumes every belief of evil as a reality. When we cease to look at Spirit, we grow blind to it, then we seek elsewhere for our good and become slaves to that which supplies us, Israel is in bondage to Egypt.

5

127

"Now there arose a new King over Egypt who knew not Joseph." Material sense and its ruling laws at first make fair promises to our thoughts that turn to dwell in it, but later on must prove a failure. It promises a consciousness of good, it gives lack of all things. Spiritual Law is the same, yesterday, today and forever. Not so man's law. Scarcely are we settled in and governed by one ruling opinion, e 'er another and a very opposite one rises up, and claims that we must serve it or die. We think of this when we hear of a new means of health, a new law to govern us, a new King ire Egypt. Under the succession of new "kings" in personal sense, new burdens have been laid upon us, one after another new laws, new fears, have arisen in the external until we serve numberless health laws and are under the greatest of slavery. We cannot sleep comfortably, we cannot eat in peace, we cannot walk, ride or sit still in safety. Yet we must sleep and eat to live. Who can live without eating? And yet, just as often eating kills us! This is certainly the Tree of the knowledge of good and evil; we are partaking of it daily and the outer made a cause enslaves us more and more.

And they, these kings, made their lives " bitter with hard bondage." Stinging remedies, bitter medicines, operations, cruel laws saying, "Disobey me and suffer." Yet, obeying, we suffer! Slaves

128

___

indeed, slaves of fashion, in food, in dress, in actions, in words; slaves of fear, afraid of everything; slaves of appetite; slaves to the body, which says feed me daily or I shall grow weak and die, clothe me or I shall freeze, fan me or I shall melt; serve me, serve me, the body constantly cries, and we are its slaves. Subject to want, to sorrow, to ills of all kinds, surely like Israel, one cry must at last go up to God by reason of our hard bondage. "And God looked upon the children of Israel and had respect unto them." Their cry found the Divine Presence at hand, ready to deliver, there is deliverance!

Return to the Top  Return to the Top

F. Moses, the Leader of Israel.. - p. 129

F. Moses was not the first discoverer of the presence of Jehovah, but it was through him that it came to be the fundamental basis of the national existence and history. From historical tradition, Moses was writer of The Torah, first six books of the Old Testament.

To a weak people, comes always a messenger outside of themselves. Those having ears, but using them not to hear; having eyes, but using them not to see! who wish to hear and   Ex. 3: 1-10 see through some one else's ears and eyes, looking only outside of themselves for the Divine message, shall never hear nor see the Truth in its full beauty.

The Truth is ready to reveal itself to each soul

129

___

as soon as it can be received, and all the teaching received from another is simply preparatory to the Divine guidance from within. Each must become strong enough to walk alone.

Jer. 31:31-35   In all ages, there have been great spiritual leaders among men, and although a necessity, it is a sign of weakness and not of strength, so it is written, "For the day cometh, when I will make a new covenant with the house of Israel; I will put my laws into their minds, and write them in their hearts. And they shall not teach every man his neighbor, saying, 'Know the Lord,' for all shall know me from the least to the

greatest."

Covenant of which Jesus was the exponent.

Heb. 8: 8-11   Moses represents the law giver who comes to that development which must be governed by something outside of itself. Such a leader does not have place in more advanced understanding.

True consciousness is not led by anything less than Absolute Truth, Impersonal Light and Wisdom received within each soul. Moses was a necessity to spiritual thought in its first emergence from bondage, and in the wilderness journey which followed.

While thought cannot stay itself upon Divine Presence and Power, because it is not realized as substantial Truth that is all and in all, it looks to find a representative of that Truth, but in doing so leans upon something less than the Perfect

130

Consciousness. There is danger both to the leader, and to those that are led by personality, or by individual opinion; to the leader lest he usurp the place of Divine authority and claims to be the guiding voice. This is illustrated in the experience of Moses.

At one time in the journey from Egypt to Canaan, we hear the Israelites murmuring against Aaron and Moses because there is no water to drink. "And the Lord spake unto Moses, saying, Speak ye unto the rock before their eyes, and it shall give forth water, so thou shalt give the congregation and their beasts drink. And Moses and Aaron gathered the congregation together before the rock and said unto them, Hear now, ye rebels; must we fetch water out of this rock?

And the Lord spake unto Moses and Aaron,

Num. 20: 10, 12

"Because ye believed me not, to sanctify me in the eyes of the children of Israel, therefore ye shall not bring this congregation unto the land which. I have given them." Ex. 9: 16

We have been slow to catch this lesson; it shows us plainly that if any individual leader claims power even to bring forth the water of life from its source, if he claims that power to be his own, individually, he dishonors Truth, because he fails to sanctify, or to hallow, the Infinite and Universal Power, which may work alike in all who turn to and obey it. For this mistake Moses never entered

131

---

the promised land; and we can now see that to claim special individual power is a false belief which must perish before the soul enters into spiritual consciousness, Canaan. The belief of individual power to bring forth the Light and Truth to the people is a mistake that belongs to less than perfect understanding. Isaiah declares, "The leaders of this people have caused them to err, and they that are led of them are destroyed."

It is claimed today that no one can go higher, or beyond, his leader. True, so we need to choose carefully what shall lead us and this lesson seems to come to us also from this history of Israel, for not only did not Moses enter Canaan, but we read

Num. 32: 11 — in the words of the Lord that "None of the men that came up out of Egypt, from twenty years old and upward, shall see the land, because they have not wholly followed me; save Caleb and Joshua, for they have wholly followed the Lord."

We see also the danger to those that are led, for they are likely to forget the True Invisible and Omnipresent Power that is in all; because of their adherence to limited power in personality they cannot come into their own possibilities.

The plea for the necessity of a personal leader is made today from this history of Israel. "We need," it is said, "a leader such as Moses was to the Israelites." Are we going backward? We have had Moses, and later one greater than

132

Moses! We have had the highest personal leader that shall ever be given to thought. In Jesus we have had our ideal leader in personality; we cannot find one beyond him, nor do we wish for one, but we shall "Grow up into him in all things," and finally awake to find ourselves "In His likeness."

Then it would be retrogression to seek or claim another leader like Moses. The day for personal leaders is drawing to its close, and the soul now seeks nothing less for its light and guidance than the Impersonal Truth, even the living Christ presence felt and known within each consciousness.

Jesus says, "Whosoever shall drink of the water I shall give him, it shall be in him a well of water springing up into everlasting life." Moses, as a leader, claimed to draw the water   John 4: 14 for the thirsty people from a source outside of themselves. He declared his power to do this and bade his followers to come and drink. This condition of consciousness and all who were led by it died without finding their land of promise. Jesus as a leader, says, "Of myself I can do nothing, the words that I speak are not mine. Why tallest thou me good, there is none good but one." He points always to the one Invisible Source and. Power, declaring to all their access to this Source; and he bids his followers to drink from the perfect satisfaction that wells up within their own souls.

133

The third and last leader is the Spirit of Truth, Impersonal and Omnipresent, guiding and leading its followers by the light within themselves, to find the Infinite Light and Fulness that belongs to each soul in the Eternal Spirit of the Universe.

Acts 10: 34

Rom. 8: 29

Israel has always stood to us as God's chosen people. This suggests to us the question, Can that Love which is no respecter of persons show any partiality? Is there any truth in the thought that God has a chosen people? Every thought of man has a truth in it, or underlying it. His interpretation of that truth may be at fault because of his limited consciousness, but this does not affect the Truth. By untruth is truth shrouded, not altered. There is a truth then in the idea of God's chosen people, although man's misconception of it has well nigh entirely shrouded it. To find the spiritual significance of "Israel" we shall need to understand the truth that is contained in the phrase, "God's chosen people." What has the Divine Love chosen above all else? What is that "Whom He did foreknow and predestinate," which means predetermine, foredoom or settle eternally?

Eph. 1: 11

Isa. 42: 1

Isa. 45: 4

Whom did this Divine Love fore-ordain according to His own will and purpose? Who are the elect in whom the Divine delighteth, as Isaiah declares, "Israel mine elect"? Whom the angels shall gather together --- from the four winds, from

134

"one end of heaven to the other"? Who are the elect of whom Peter seeks, "elect according to the foreknowlege of God"? Whom does John address as the "elect lady"? What true meaning is there in this idea of election and predestination, brought out plainly by Isaiah, Paul, Peter, John and Jesus? Whom has the Lord chosen?

Matt. 24: 31

1 Pet. 1: 2

2 John 1: 1

Setting aside the thought of personality, we find a satisfactory answer to this question. When God is recognized as changeless principle, the Lave, Truth, Life and Being of each individual, and because Omnipresent, present alike in all and to all; when Truth is enthroned in our consciousness and begins its work of judgment within us, separating Divine Nature from all that claims to be material, or mortal; when we can hear Truth's decision saying to Divine Nature, "the sheep," spiritual nature within us, "Enter thou into the joy of thy Lord," also saying to all claims of mortality, "the goats" within our thoughts, "Depart from me, I never knew you"; when this light of consciousness has shined within us; then do we understand what or whom it is that has been eternally chosen by Truth and Love. Divine Nature is the elect of God. It belongs to every living soul in God's universe. It is the Life and Substance of Spirit in each, predestined by Divine Will and Purpose, to come out at last pure and 5a

135

---

undefiled from all claims of mortal sense which Truth never knew.

Blessed is the assurance of predestination, I cannot be eternally lost you cannot be. "Ye have not chosen me, but I have chosen you," is said to each soul forever. "I stand at the door and knock." I do not knock and go away if you will not "let me in," as our once beloved hymn declares --- "If you wait he will depart." No! I stand until you open. I am Omnipresence, whether you recognize Me or not, saith Spirit. When you have acknowledged only Me, you have "opened the door," you have recognized yourself

as the chosen. "For this cause have I raised thee up for to shine in thee my power, and that My name may be declared throughout the earth."

Ex. 9: 16

In Jesus' parable of the wedding supper, we are shown who the "chosen" are. When those who were especially bidden would not come, did not "choose" to come, the king said to his servants, "Go into the highways and as many as ye shall find, bid to the wedding." So the servants went out and gathered together all, as many as they found, both good and bad, and the wedding was furnished with guests. "For many (all) are called, but few are chosen." He who chooses to come, is chosen. Only the Divine in man will choose to enter in.

In that wonderful Divine Name I Am, each

136

soul is called. My new consciousness of what I am is my new name, I Am is the family name, the the Name above every name at which every knee shall bow, the name for whom the whole family in heaven and earth is named. We know this wonderful name, Nature Divine, when we have come up over all claims of limitation in the earth or world beliefs and opinions.

"To him that overcometh I will give a white stone (pure truth) and in the stone a new name written, which no man knoweth saving he that receiveth it." Truth gives us a new consciousness of our nature, seen in the Divine I Am. Divine Name or Nature belongs to all, for "Have we not all one Father, one Source, one Life, one Substance, one Nature? Israel, then, is the elect of God, is that which God has chosen, for Israel is type of that which is spiritual in each. This is the Truth that was partially heard by Abraham, "I will make thee a great nation," and interpreted as a special or individual promise. The fuller meaning came later to Abraham, " Thou shalt be a father of many nations." It was heard more clearly by Jacob, "I will make of thee a nation and a company of nations."

Rev. 2: 17

Isa. 45: 4

Eze. 20: 5

Gen. 12: 2

Gen. 35: 11

To the limited conception the God of Israel has meant the God of the nation Israel; in the broader truth it means the God of the nature of Israel, the God of all spiritual nature. So Paul declares,

Gen. 17: 4

137

"Know thee therefore that they which are of faith the same are the children of Abraham. And if ye be Christ's, then are ye Abraham's seed and heirs according to the promise." God has chosen all who choose God.

Gal. 3: 7

Gal. 3: 29

In one way or another, all choose God, for there is something in all choosing Good, even though feebly. No one is utterly devoid of good feeling. This something in all, that shows out at times, which times we call their best moments --- is proof that in each is some spark of God nature. This little something of good, or remnant, is the Israel, the elect, chosen, ordained to be sooner or later conformed to the image of His Son. It is that which in the "beginning" was made in the Divine "image and likeness."

It is written, "So all Israel shall be saved." Israel then represents to us the spiritual thought or nature, born of God in each. In studying the history of Israel as given in the Old Testament we may say with Paul, "Which things are an allegory," and find in the circumstances as related a figure applicable to any one in his first feeble consciousness of Spirit and of the relation of self to Spirit. Mount Sinai, Mount of Spiritual Wisdom, disturbance on Mount," confused mental condition of people below." Moses received the decalogue through

Rom. 11: 26

intuition of the voice within. Angel of

138

Lord appearing in burning bush, his higher realization; voice of God, his inner conviction.

"Son of Man, set thy face against Pharaoh, King of Egypt, and against all Egypt," Truth is opposed to all personal beliefs, because the belief in Cause as originating in what is called materiality, is adverse or contrary to Truth.

"Behold, I am against thee, Pharaoh, which saith my river is mine own (my thought and opinion of Truth is my own, man-made creed) and I have made it for myself. All the inhabitants of Egypt shall know that I Am the Lord --- and the land of Egypt (the belief in anything apart from Spirit) shall be desolate and waste --- they have been a staff or reed to the house of Israel (Spiritual sense not yet conscious of perfection); when they leaned upon thee thou brakest. Behold, I am against thee and thy rivers, and it shall be no more the confidence of the house of Israel; but they shall know that I Am the Lord God."

Eze. 29: 2,
3, 6, 9,
10, 16

Return to the Top   Return to the Top

G. Death of the First-born and the Passover.. - p. 139

G. The last touch that the finger of destroying Truth placed upon "Egypt" was the death of its first born. This reached the very heart and core of material claims, hence it came last.

Ex. 11: 4, 5

The "first-born" is the starting point of our belief, hence it is the cause and foundation of all

Ex. 12: 13,
21-23, 29, 31,
51

139

---

that is claimed in that belief. To strike at the "first-born," is to reach and annihilate the very basis or reason for any material claim.

The "first-born" of material sense is the claim of Life, Substance and Power in a material source. That we have Life, Substance and Power from the flesh, is the "first-born" claim that originates every other fleshly claim.

The death of this "first-born" then, is the destruction of the very foundation of misapprehension, or belief in materiality. This destruction does not touch Israel, for it has place only in material sense, but it is the final stroke that frees Israel from the claim of bondage to the external as cause.

Israel is shown how the angel of death shall strike the Egyptian household, but shall pass over the houses of the Israelites, and in faith they prove their understanding of this by celebrating the Passover, before it has taken place! So our consciousness, trusting in Spirit, sees and celebrates in its thought the understanding that destruction and death do not touch anything Spiritual, but, like an angel of mercy, destroy all those claims which hold us in captivity. With equal faith, we celebrate this truth before we have seen the Passover take place. We can rejoice in Truth revealed to our consciousness even before it is demonstrated to our eyes. Faith, the evidence of

140

---

things not seen, makes us sing for joy before there is any outer proof.

The slowness with which we come out from under the burden of material bondage is well represented by the stubbornness of Pharaoh in letting Israel go. Not until the "judgment of the Lord" falls upon the dearest idols of material belief does the hold of materiality upon Spiritual thought relax.

Clearer light on Truth gives thought clearer judgment; by this light thought begins to see the true nature of that which has claimed its service, and to understand the powerlessness of that which claimed great power. By this consciousness which discerns more truly than the former, false claims are separated from Truth and the power of such claims is broken.

After the ten plagues have fallen heavily upon Egypt, Pharaoh bids Israel go; so, after all our idols of material good have fallen, the material lets go its hold upon us, or to put it more truly, we are able to break away from materiality, which no longer controls our thought.

Are we, from this point, free? We might be if we would!

The going of Israel out of bondage to Egypt may be compared to the flood, in which the world of the ungodly perished; for both events may stand to us as types of that stage in unfoldment

141

---

when false claims are destroyed to Spiritual thought, by the incoming of fuller light (of understanding) upon the truth of all things.

So we may ask, why are not all false claims forever silenced when the floodlight of Truth has 'purified our thoughts, or when our Spiritual thoughts have left the "land of Egypt" (materiality) In the first case all claims of Being besides God perish; in the latter all idols of material sense are shattered. Why is not this the end of our struggle with material and fleshly claims? Because this being a process in the purification of thought, the true vision is first seen in the recognition of the destruction of all misconceptions in the Presence and Power of Truth; thought then unfolds into realization of this true vision.

The fact that our highest thought has seen. Truth and knows upon what to base its understanding, "Holy Ground," should make the thoughts that follow press on faithfully in order to enter fully into that which has been revealed as truth. All hindrance to steady and rapid growth into consciousness, is because thought wavers. It does not turn a deaf ear to the protestations of old claims even after it has seen their destruction in Truth's Presence. So long as thought listens to the claim of physical causation there is a threatening power in it. The method by which Israel is freed from Pharaoh's power is the method still by

142

---

which thought is freed from material claims that rule it.

Israel obediently followed Moses out of Egypt. Let our spiritual thought obediently follow our   Ex. 14: 2, 4, 10,
                                               13-16, 19-20,
highest consciousness that leads out of bondage.   22, 23, 26, 27

There is heard the command, "Sanctify unto Me,   Ex. 13: 2, 21-22

all the first-born --- it is mine." Let the reason, or starting point, the very beginning of your understanding concerning Divine Manifestation, be in the I Am that I Am. Reason of Truth from consciousness of the Divine Me, God, the Source and Cause, All expression is like its source, "like begets like." That thought does not see Divine leading clearly is hinted at by the fact that it is said, " The Lord went before them by day in the pillar of fire." These represent symbols of Divine Presence, instead of a realization of the actual Presence; and all symbols are the result of human belief, a concession to human weakness.

Rev. 21: 1

Soon after the departure from Egypt, Israel "encamps by the sea." What is this "sea"
before which our thought, seeking spiritual consciousness, halts I In true consciousness
there is we read "no more sea." Then the sea is something met with in undeveloped consciousness of spirit!
It is something that hinders Israel's march to Canaan, and before which Israel hesitates. It must indicate
some hindrance which we hold in our own thought, some belief of separation

143

from Spirit and of our limited power and possibility. This is the conception before which we each have halted after seeing our freedom in Spirit. A shadow crosses our path! To be led by a "cloud" and " a pillar of fire" instead of by realization that God's Presence is all, is to be subject to shadows. A realization less than the Presence of full light is sure to see shadows.

The very belief in the necessity of taking a long journey to reach Canaan must have within it the claim of of separation from the Divine, hence a lack of understanding of our possibilities. This is the sea that impedes our progress until Divine Light shows us how to pass it. But lack of consciousness pauses here, and while it hesitates, the host of Pharaoh is seen coming to overtake. Every hesitation in the direct path of Light and Truth brings upon us claims from the past. In true consciousness, " Time shall be no more," then there will be no past claim recognized but Israel has not yet entered full consciousness and the old material claims have not been left very far behind! They are yet remembered.

" They were sore afraid." "Perfect love (which is perfect consciousness) casts out all fear." This fear is another proof that Israel is not yet perfect in consciousness the journey in the wilderness has no place in perfect consciousness. Israel does not yet know, when the old claims

144

threaten, as those of which we have not yet lost memory will do, that their power to enslave is gone. Divine Love must give still greater proof of this.

"And Moses said unto the people, 'Fear ye not; stand still, and see the salvation of God ---
for the Egyptians whom ye have seen today, ye shall see them no more forever. The Lord   Ex. 14: 13
shall fight for you and ye shall hold your peace.' " We are all familiar with this point in our
new consciousness, our new attitude toward Life. After we think our bondage to sin, sickness and death has ended and that we have stepped into great freedom of thought, we are sometimes suddenly stopped by some condition that claims, " You cannot ignore me, for I prove to you that your possibilities are limited; that you are not one with Almighty Power and Goodness; you are but a mortal being after all your great claims." Has there not one thought been arrested in its flight from material bondage to acknowledge that, perhaps, after all, we do have to serve the material, and have been mistaken in supposing ourselves free?

This, then, is our sea. Here are we about to be overtaken by Pharaoh's host and almost discouraged, all because, when we saw a claim of separation and limitation in the so-called flesh, we allowed our thoughts to falter, to waver, to doubt! The highest consciousness within us meets our

145

doubt if we will listen to it. "Fear not, stand still; see the way in which the Divine saves." This is the first assurance that quiets thought that it may hear the new command. In this emergency, our highest thought turns and cries to the Divine and a strange answer comes: "And the Lord said unto Moses, 'Wherefore criest thou unto me Speak unto the children of Israel that they go forward,' " as if to say, "I have opened the way by giving you the understanding how to leave bondage; I can do no more for you, so do not ask Me for help, but do as I have told you --- go forward, push on, let no material claims overtake you -- progress out of their reach." "Lift up thy rod and stretch out thine hand over the sea and divide it, and the children of Israel shall go on dry ground through the midst of the sea."

"Lift up thy rod and stretch out thy hand over the sea" let the thought of thy power, the Truth over all claims of limitation, be exalted in your consciousness, then shall you be able to walk through all claims of separation and lack of power upon the knowledge of a firm and fixed foundation; you shall not then lose your "footing," you will know whereon you stand, you will be able to walk through, or progress right on in spite of seeming limitations, because conscious of your true power and certain of the ground of your faith.

146

In this night of waiting, the "pillar" that was light to Israel was clouds and darkness to Pharaoh's host.

"And the children of Israel went into the midst of the sea, upon dry ground, and the Egyptians pursued and went in after them to the midst of the sea." Now, indeed, is Israel   Ex. 14: 27 brought to the severest test of faith, for even after obedience to Divine command, having passed right into the midst of the sea on each side, holding to their true foundation, still they see the relentless pursuit of old. enemies! There is yet something more for Spiritual thought to do. " Stretch out thine hand (again) over the sea --- and the Lord overthrew the Egyptians in the midst of the sea."

If Israel had once faltered here, their hope would have fled. Again must the strong confidence in their God-given power and possibility be lifted up in their own consciousness, if they would see the final destruction of the claims that held them in bondage for so long.

Pharaoh's host following, even upon the dry ground of the sea through which Israel has passed, shows how persistently our thought turns back to find ignorant beliefs at its very heels, even trying to attend us when we have found the ground for our perfect faith in Divine Power.

If our thought will not fear, but will hold persistently to our foundation for true knowledge, all

147

---

false claims will find their grave upon this very "dry ground" that has borne us safely through.

Return to the Top   Return to the Top

H. Deliverance from Egypt.. - p. 148

H. In the bondage of Israel to Egypt we find perfect type of the enslavement of our thought when it wanders from consciousness of Spirit as All and chooses the material as its helper. In the Egyptians we trace a likeness to those material claims of good and ill which sway our thought while it believes in separation of body from Spirit it believes that while we are "in the body we are absent from the Lord."

Canaan --- Spirit. Egypt --- body.

Bondage in Egypt is when our body has become our master, because thought has claimed it to be a cause and believes in its separation from or unlikeness to Spirit. To be delivered from this bondage to ignorance, thought must learn the truth from its foundation. It must be taught that the very basis of all life is holy and that, therefore, all manifestation or form of life is holy. Thought is to see our dependence upon Spirit, and not to conceive of dependence upon body. It is to see Spirit as Cause and body as effect. The whole body, earth, or external, is the result or effect of Spirit, emanating from within Spirit.

The external is not Cause, but Spirit made visible. And thought is to know the unity of Cause

148

---

and effect; if cause is good, effect is good. It is to see one and the very same life, substance and perfection for both Spirit and body, but it must hold Cause as Cause, and effect as effect.

The very first words of deliverance that came to Israel after their cry went forth were these words to Moses: "The ground whereon thou standest is holy ground." This teaches the Omnipresence of Cause. The "ground" is the foundation, and the message is, "Find right here and now that the ground whereon thou standest is holy."

"The kingdom of heaven is at hand, the kingdom of God is within you," right here while you stand in the body.

Oh, that Israel could have understood this Truth; but Israel today does not hear it; she stands ready to leave at the summons of Truth and to take the "long journey" to the better land. As Israel leaves Egypt, seeing the kingdom, the holy land, the holy ground "afar off," she wanders in the wilderness of her own making for forty years.

Moses is sent to the King with the message, "Let my people go." Moses is the type of the highest consciousness of Spirit that has come to us in our bondage. It says within us, "all these laws you are serving have no real power over you -- break the yoke and go free."

Pharaoh represents the strong-hold which these laws that we have learned to serve, have over our

149

---

thoughts. Pharaoh's obstinacy and reluctance to let the people go, give us a picture of how firmly these material claims have taken possession of our thoughts. We cannot throw them off without a desperate struggle. It seems impossible to know ourselves free from such long recognized authority, the authority of Man over Man.

Our own conceptions of power and wisdom in personality is the Pharaoh that says, "No, I cannot let them go." Then the Lord says, "See what I shall do --- for with a strong hand shall he let them go, shall he drive them out of his land."

Ex. 12: 12 Suffering shall drive if Love does not draw. To us first comes a dawning light of the Way, the Truth and the Life. If we answer, "I cannot follow I cannot let go," then Love says, "See what I shall do." The light that would lead us becomes a consuming fire, destroying that which has held us. Unconsciously we find our dearest ideas of good being reversed by Truth's power and presence, that, in our opinions and beliefs, which has blinded us to seeing God as All, begins to yield to higher power. Our idols fall because in our clearer light we see no good in them. Thus is that Strong Hand laid upon us. "I will bring My people out of Egypt by great .` judgments,' separating by My Divine Light the true from the false in their consciousness." These judgments are typified by the ten plagues. "Against all the gods of Egypt will

---

I execute judgment"; against all the good that the material sense claims. One says, "In each of the plagues of Egypt, some one idol of the Egyptians is struck at." This is suggestive that when Love cannot draw us from our false ideas of good, it drives us by destroying that claim of good.

We need not be surprised, after more light has come, that we cannot go back and find the same satisfaction in our old forms of service, or in our old beliefs of life. It is Truth in its   Isa. 29: 13-14 work of destroying our many loves that we may find all in the one changeless love.

The ten plagues drive us to seek wisdom in something more substantial and enduring than man's opinions of Truth. Each individual must   Luke 17: 21

learn not to draw from many sources to perfect his consciousness, but to draw from, or be sustained entirely by, the one source within. " The kingdom of God is within you." The plagues come to us when the external begins to lose its attractiveness to our thought and the inner or spiritual seems the only desirable possession. Then, that which has tempted thought to turn from Spirit, is given up, and we start on our search for the better thing.

The slowness with which we emerge from under the shackles of external claims is further illustrated to us in the long wilderness journey. As Israel obediently followed Moses, so must our

---

spiritual thoughts be obedient to our highest light. Moses is the leading thought in that consciousness which says, "Heaven is ours." It is that distant land towards which we journey, our ground of faith is holiness which we must keep in sight while we journey. We leave the body, or belief in body as cause, to take possession of all spiritual good, and so the journey begins.

In the increased light of our understanding today, we see that there is no journey to take where unity of heaven and earth, of soul and body, is seen. All hindrance to immediate acceptance of heavenly conditions, is because thought wavers "between two opinions" not yet seeing clearly on the one hand, and not giving up wholly on the other. It does not turn a deaf ear to old claims even after it has felt Truth's destroying power.

That Israel did not see clearly the Divine Presence is indicated in the words, "The Lord went before them by day in the pillar of a cloud, and by night in a, pillar of fire." These are but symbols of that Presence. Symbols are props to the weakness of human thought. We feed upon the symbol while we do not realize the Presence of the reality.

1 Cor. 13: 10   We cannot use the symbol after the Presence that is with us always has come to our consciousness "Do this till I come." "When that which is perfect is come, that which was in part shall be done away."

The "sea" before which Israel halts soon after leaving Egypt is the belief of separation between Canaan and Egypt, between Spirit and body. This "sea" is some obstacle to progress or to the onward march of the soul while consciousness is not fully enlightened. While Omnipresence is not realized, shadows must cross our path. We yet sigh for the flesh pots of Egypt. To lose sight of the ground of our faith is to halt by the "sea." There we take a backward glance, and fast coming upon us is seen the host of our old enemies. This is a severe testing time for the new consciousness.

But one direction comes from the Divine side, "Fear ye not. Stand still and see the salvation of God. The Lord shall fight for you and ye shall hold your peace." Every hesitation in the on ward path brings into our memory old claims of fear and bondage. To " stand still and hold your peace" is to be positive and firm in your consciousness of God's Presence and Power. So the command follows, "Go forward," "Lift thy rod and stretch thy hand over the sea." Conquer all that says "Impossible" by your knowledge of the true Source of Power. To stretch thy rod over the sea is to deny separation from Spirit. Then the waters of the sea roll back; we see again our solid ground, and walk through the "sea" upon dry ground.

Our false conceptions do not taint the purity

153

of our real life. I Am what I Am, despite all false beliefs. When every claim of separation is destroyed, then we read in Revelation, " There is no more sea."

Pharaoh's host attempt to cross on the same ground over which Israel passed. Death claims to be as real as life. Evil --- ignorance, makes the same claim to true foundation as Good. Ignorance imitates Truth, but when once .spiritual thought is founded upon true consciousness, all claims opposite to Good find their death in our consciousness upon this very ground which sustains all Truth.

The "ground," or reason, we have found in God is also our ground for detecting the false. On the same "ground" that we know what truth is, we detect what ignorance is --- and to know ignorance as ignorance is its destruction. When we have crossed over the claim of our separation from Spirit we see our enemies, sin, sickness and death, forever buried from our view under the power of our faith.

Divine Being as All and in All is the ground of our faith; upon this stand, be firm.

Do we believe in Truth for what it gives us; or do we follow for love of Truth? Would we rather die with it if need be, than live without it? If this is our consciousness, happy are we, for then our thought will not waver nor be diverted by outward appearance, but we will say with Paul,

154

"None of these things move me"; or with Jesus. "The prince of this world cometh and hath nothing in me."

Return to the Top Return to the Top

I. Entering Canaan.. - p. 155

I. "For bodily exercise profiteth little, but godliness is profitable unto all things, having the promise of the Life that now is, and of that which is to come."

So we see they could not enter in because of unbelief. "All things are possible to him that believeth."

Let him that is on the housetop not come down to take anything out." "House" is old habits of thought. From the housetop eternal truths are revealed. Let him who receives the vision leave all things connected with his old habit of thought.

Israel's delay in entering Canaan was occasioned by their own lack of faith in God. Believing in separation,

"These all died in the faith, not having received the promises, but having seen them afar off. For they that say such things, declare plainly that they seek a country (have not yet found true consciousness), and truly if they had been mindful of that from whence they came out, they might have had opportunity to have returned." Sighing for the flesh pots of Egypt, they could not take hold of Spirit's fulness with unwavering confidence. In this feeling of lack,

155

they cried out, "Would to God we had died in the land of Egypt, where we sat by the flesh pots and where we did eat flesh to the full, for ye have brought us out to kill us with hunger." They forgot the rock of their salvation in the Eternal they lost sight of the ground of their faith. So the decree was pronounced, "Ye shall not come into the land because ye are turned away from the Lord, the Lord shall not be with you. Surely none of the men that came up out of Egypt shall see the land, because they have not wholly followed Me.

Num. 14: 31

But your little ones --- them will I bring in, and they shall know the land which ye have despised."

"Within one year after leaving Egypt, Israel draws near to Canaan," yet they did not enter into possession of the land.

Opinions and beliefs within us that insist upon their own knowledge, never carry us into Spiritual consciousness. Such condition of thought dies in the wilderness that truer thoughts may realize the Eternal. Ignorance is not a passport to heaven. To know God and God's manifestation is Eternal Life.

Deut. 18: 13

Matt. 5: 48

In Divine Science, the starting point is the unit or One God. All forms are the expressions of the One; they exist as the One revealed. Each living form is as perfect within itself as is the One producing it. "Be ye therefore perfect, even as your Father which is in heaven is perfect."

156

"Existence is not here on its own account, but is here because God, the Creator, is here. All Cause must show forth in effect. The Creator, or Source, must appear in the creature."

We worship God in Spirit and in Truth when we believe in God only. Since God is infinite, He demands infinite worship. When we give God this infinite worship, we acknowledge God as all there is and cannot believe in any opposite reality.

"He that stoppeth his ears from hearing of blood and shutteth his eyes from seeing evil, he shall dwell on high he shall behold the land that is very far off."

Isa. 33:15-17

"If ye be willing and obedient, ye shall eat the good of the land, but if ye refuse and rebel, ye shall be devoured with the sword."

Isa. 1: 19-20

Hos. 11: 1

"Out of Egypt have I called my Son."

157

Return to the Top Return to the Top

CHAPTER 3.

EXODUS.. - p. 158

Moses was forty years of age when he saw the burning bush, while feeding Jethro's flock in the solitude. Human individuality in its conscious unity with the "consuming fire" of God's presence, the universal Spirit of Life. We have all seen the burning bush and heard the voice.

We know when we understand the Truth that the inner and outer are all sacred.

Adam is not ignorance or error, and does not stand for ignorance or error the child is just as perfect as a child as an adult is as an adult.

"Feet" means understanding. "The place whereon thou standest is holy ground." If we knew that, we should not try to better ourselves by going to another place; if we were certain that there was no place in the universe where there was more of God, how that would change things for us.

Evidently, Moses did not understand this, for he went on. However, Moses knew his Divine possibilities as no other man before him had known. The law of Sinai was given the fiftieth day after leaving Egypt. The Tabernacle was planned after the vision of Moses upon the Mount. We

158

believe that in those days every direction that came to the Hebrews was supposed by them to be the voice of God. That was how they heard the commandments to, do these different things; that was why they were afterward contradicted. All these directions about the temple referred to man's own life, but Moses took them as directions to build an external temple; "he built it according to the pattern shown him on the Mount." He did faithfully as he understood. In Revelation it says, "I see no temple there." It was wiped out in the highest consciousness. Rev. 4:1-10. Personal sense in Moses argued. Even by Moses' conceptions, he could not change his divine nature; wrong conception does not know the law.

It was personality that kept Moses from leading the children into the promised land. He said, "Let us bring the water." No person is ever going to lead people into the Truth. "Know no man after the flesh." Not th4 personal Jesus, but the Christ within will lead us to the promised land. This people did not expect to hear the voice of God. They expected Moses to hear it and tell them what it said! Our "Moses" stands for our leading vision of our present light; type of his vision in his day (light) as leader.

Pharaoh went back into his house --- old habit of thought. Pharaoh is the ruling thought in the realm of belief. Nothing that makes a claim.

Ex. 5:2, 7, 23

6

Ps. 2: 1-4

159

Heathen belief. "Hardened his heart." Pharaoh was harder to impress after his persistent refusal.

Ex. 9: 27

Ex. 6: 29-30

Pharaoh progressed. "Death of first born," death of our dearest conceptions. We cannot hold a belief that is a betrayer of truth. We must look to the end of our conceptions sooner or later, the right or constructive way. Aaron is given Moses for mouthpiece.

Ex. 7: 1

Aaron's rod swallowed the others; ignorance is swallowed up by Truth; Truth is the only power.

Third plague magicians could not help, "it is the finger of God."

Ten plagues stand for the torments which come to us. That which has been called possibilities in the external fail us, we meet disappointments.

Ex. 10: 26

Moses would not go free without the animals.

Ex. 14:19-20

Cloud is Omnipresence; it was the spiritual light, always present.

Ps. 34:7

Moses' song. Epic poetry. Read it.

Ex. 15:

Ex. 16:

Obstacle --- the sea; instead of going forward through the barrier, they halted and soon were overtaken by their old foes. Sea, here, stands for separation. When we place anything between ourselves and the presence of God, our old conceptions and beliefs overtake us.

Israel does not represent a very high plane of consciousness.

"The Lord shall fight for you," and "Lift up

160

---

thy rod and stretch out thy hand"; we have to do both; go to our source and let the Omnipresence direct and correct our old beliefs.

"Rod" stands for correction or discipline, individual.

Ex. 17: 8-9

"Thy staff" stands for the Truth. Thy Truth comforts me.

Ex. 15:22-24

Wilderness. Found no water in wilderness."Water," divine possibilities. When we are not in our spiritual development, we do not see our divine possibilities.

Ex. 16: 3

Hunger. Quail and manna the answer. Manna fell six days, and had to be gathered each day except the seventh. There comes a time when we do not have to gather. Children of Israel "having seen these things afar off, died in the faith." We are seeing them in our midst; that is what the consciousness of the Omnipresence is doing for us. We should go to our Source every day and gather our Spiritual Manna; getting greater realization of the Presence of God.

Ex. 16: 12-21

Ex. 16: 30-35

No water. We must thirst for the Truth. "Struck the rock," correct our old mistakes and beliefs.

Ex. 17:1

The remedy! Second Commandment. "Visiting the iniquity of the fathers upon the children unto the third and fourth generation of them that hate me."

Ex. 17: 5, 6, 11-12

161

Ex. 20: 5-6

Ezekiel, eighteenth chapter, blots out heredity. Love is consciousness; hate is unconsciousness.

Moses' law.

Ex. 21: 24-25

Lev. 19: 18   Jehovah speaks unto Moses.

Matt. 5:44-48   Jesus' teaching. Two commandments. Mark 12:29-31, and read 32-35.

Ex. 23: 25   Divine Healing.

Ex. 24: 13   Joshua, minister of Moses.

Joshua was beginning to have a higher vision than Moses had; he was more trusting; he led the children of Israel into Canaan. He was a follower of Moses but went beyond.

The glory of the Lord is always on the mountain (clear consciousness). Moses had to stay on the mountain until his highest vision was realized.

Ex. 25:   First step toward building of temple. The Ark was a national not merely a tribal sanctuary and its loss to them was a national calamity. "Ark" stands for the presence of God in sanctuary, a symbol. Symbols are concessions to our ignorance.

Ex. 25: 22   First communion. Covered or made of pure gold. God's presence is pure. Purity is unadulterated or unmixed.

Ex. 25: 40   "Pattern," God's idea of perfection when on the Mount.

Ex. 28: 15   "Perfect body" is holy garment of Aaron the spirit of God must be clothed perfectly. Breastplate of judgment is essential to the priestly

162

---

thought or holiest thought. Divine nature in us is the high priest within us.

Oil is consecration to pure consciousness of what Life is. Two tables written by Moses; duty to God and duty to man.   Ex. 29: 7

Consciousness of God's presence is always rest. Moses was on the Mount forty days and forty nights.   Ex. 33: 14

     Ex. 40: 3

``Veil" stands for belief of separation. With each revelation,-the veil grows thinner. Consecration of the tabernacle is the consecration of our body.

The reality of the cloud is God's presence. Ex. 40: 38

When cloudy vision of the body is lifted up, we progress.

"Night," deep things of Truth.

Moses says, " Of my brethren shall the Lord, thy God, raise up a prophet like unto me"; the prophet that was to come should continue his work and when that prophet came, he said, "If they hear not Moses and the prophets, neither will they be persuaded, though one rose from the dead." Jesus came to fulfill the law, not to destroy it.

163

CHAPTER 4.

LEVITICUS. . - p. 164

Leviticus and Numbers

THE BOOK OF LEVITICUS.

Describes the ceremonial law.

Sacrifices. There were two kinds of sacrifices; one from the fruit of the earth, supposed to be made by one who did not need repentance. The other was a living animal and was always a confession of sin.

The Passover was the first sacrament instituted. A lamb without blemish was taken, the blood scattered and the flesh eaten with great joy. This was the last feast made in Egypt; henceforth they were to depend upon God for their food.

" Sacrifice " --- making sacred. We have thought sacrifice meant self-denial, but it is not so.

Lev. 1:3,	--- Sacrifice Burnt offering; of cattle.
Lev. 2,	--- Sacrifice Meal; of herb.
Lev. 3,	--- Sacrifice Peace; of animals.
Lev. 4,	--- Sacrifice Sin; of bullocks.
Lev. 5,	--- Sacrifice Trespass sin that is known.

Rom. 12: 1	Sin has no relation to good. We must present our living bodies before God without blemish.
Eph. 5: 26-7	Read the following, and see why the prophets

164

thought they were so fully directed about sacrifices:

Jer. 6:20   Heb. 9:10

Jer. 7:21-23	Heb. 10:1-9
I Sam. 15:3	Isa. 1:11-15
Ps. 40:6	Isa. 66:3

We see, as time advances, that these things which are told at early stages of man's development as being sacred, are wiped out in later writings. We have a higher call than bloody sacrifices, man's conception of what is Truth.

Peter's vision. By "unclean" is meant, hurtful. To Peter's consciousness it came that. nothing is unclean unless we think it unclean. Animals were saved with Noah and with Moses.

Rom. 14: 14

Acts 10:11-16

"Goat" atonement; since there was no killing, a step in advance of sacrifices. "Feasts " --- the Sabbath, the Passover, fruits, Pentecost, etc.

Lev. 16:20-21

Feasts are symbols. Passover from negative to positive conditions. We are going through the "Passover" daily. Prov. 15:15, Isa. 1:13-14.

Lev. 23:

1 Cor. 5: 8

A "day" is a period of development or understanding. In the great unfoldment, six periods to labor, to attain, then a great period of rest. We are claiming our possessions which God has given us in the beginning; opening into the consciousness of Truth.

Lev. 25: 3-4

165

Lev. 23:18-19

John 8: 35

John 15: 15

Rev. 21: 3

Lev. 26:36-37

Lev. 26:44-45

Getting back to Eden.

"Servants" God calling them servants, indicated their low plane of development or knowledge.

A promise when they shall outgrow their development as servants into consciousness of sonship or realization. Our enemies are our wrong beliefs when they fall by the sword, we shall get back to Eden consciousness.

Our withdrawal from God causes fear.

"I will remember the covenant of their ancestors."

Return to the Top Return to the Top

NUMBERS.. - p. 166

THE BOOK OF NUMBERS.

God does not fight for numbers.

Num. 1: 1-3

2 Sam. 24:1-3, 10	Supposed to have been a punishment to David for numbering the people. Joab tried to show him it was not according to the Lord's will to number is to depart from our principle of supply, fullness.
Num. 3: 6-7	
Num. 6: 22-27	"Levites," priests to do service of the tabernacle. Levites, priestly nature within us, serve the Lord.
Num. 10:	Aaron and his sons gave this blessing to the children of Israel.
Num. 12: 13	Leaving Sinai, law of cloud. Ceased from their journey when the cloud lifted they went on. If we are clouded in our thought, we stand still.

Reference to Divine Healing.

The Rock which is Christ; waters, possibilities

166

---

of Christ. Speak the possibilities of Christ, "Christ in you, your hope of glory." I Cor. 10:1-4.

Num. 20:8-10

Visions were clouded, and all passed through the sea, sense of separation from God, duality. To them Moses was the highest.

John 6: 35

Amos 8: 11

The famine was not a famine 'of thirst, but of hearing the words of the Lord. The "fiery serpent" --- purifying wisdom. The people were getting into a negative condition, and murmured. "Fire" --- burning away, God's love a consuming fire. The only cure of adultery and idolatry is Wisdom Light. The fiery serpent is the power that leads us away from ignorance. The literal makes God a most cruel master, but he is always exercising wisdom to lead us out of ignorance, the thing that makes it sting is the holding on to one's conceptions. Moses made a serpent of brass; they had to look high to get wisdom.

Num. 21: 6

Num. 22:

The Moabites were distressed when they knew the Israelites had come against them. Silver and gold was a temptation.

To Balaam the Lord said, "Do not this," then later he heard God say "Go." Sometimes when there is a great temptation, we may think we hear the inner voice (God's will) to agree with our own wishes. 2 Peter 2:15-16, and Jude, 11th verse.

Greed and reward took Balaam. We cannot do

6a

167

---

things for reward. Balaam's eyes were opened by the Lord, so he saw.

	God sees only the light of us.
Num. 23: 8, 10, 11, 19-21	There runs all through the Bible the promise of a deliverer coming. This points to the Christ, of Jesus, and of us.
Num. 24:17-19	
	God's anger, when they did wrong. God's law began to work and purify.
Num. 28: 23-24	
	Sacrifice is not of God's command. They come to the border now, after forty years. All of Israel that had left Egypt are dead except Joshua and Caleb.

Num. 32: 20-23, 25-27

War is war of our false conceptions. In sight of the Lord, after they had seen the Law, yet they do evil still. "I purpose that my lips shall not transgress."

Num. 33:38, 39

Aaron and Moses both died on the Mount. Personality dies in the higher vision.

Num. 33:52, 53, 55

"Inhabitants in the land" --personal belief, belief in anything or any personality separate and apart from God as a reality.

Num. 35: 6-7

Levites were given forty-eight cities, six special ones cities of Refuge for those who had done wrong, where they could go and be protected until they had a fair hearing.

168

Return to the Top  Return to the Top

CHAPTER 5.

DEUTERONOMY.. - p. 169

Comparison of laws in Deuteronomy. Expansion of Ex. 20 and 23. Parallel to Lev. 17 and 26.

Law rehearsed, purpose of book, exhortation. Writer neither historian nor jurist, but religious teacher.

Characteristic phrases, "A mighty hand and a stretched out arm" "The land whither thou goest into possess it" "With all your heart and with all your soul" " That it may be well with thee""A peculiar people" " To make his name to dwell there"; " To walk in the ways of Jehovah"; " To hearken to the voice."

Moses rehearses the journeys of the children of Israel from the time they were rescued from Egypt until reaching Canaan.

Perhaps Moses said "For your sake," said it humbly.

Moses' prayer Moses would only see Canaan. Jealous in that "He will not let anything remain but Himself." Love, the true spring of all human action.

Deut. 1: 37-38

Deut. 3: 24-28

Deut. 4: 24, 29,35, 39

Deut. 5:10	Deut. 11:1, 13, 22
Deut. 6:5	Deut. 13:8
Deut. 7:8	Deut. 19:9
Dent. 10:12-17	Deut, 30:6, 16, 20

169

Deut. 5: 24	The Christ in us can see God and live; false beliefs are killed. Jesus got to where there were no more to kill, the ascension.
Deut. 6: 4-5,7, 9-12, 17-19	Reminds us of Jesus' second commandment. In the midst of good things, remember God consciousness that God is the Source of supply. Dent. 8:17-18.
Deut. 8: 3-4, 11-13	They are getting back to know that God giveth all.
Deut. 10: 9-12, 18	Highest appeal to consciousness of God's love, in that He had chosen Israel.
Deut. 11: 10, 18	Levites are set aside, priesthood. The priesthood is within ourselves and is the highest thought within us. Levites represent the true attitude of the soul. The highest spiritual vision is direct inheritance from God.
Deut. 12: 23	"Thou sowedst thy seed in Egypt."
Deut. 15: 1-12	"For the blood is the life." We are saved by giving forth of this life not the blood of Jesus, but blood of Life as it flows the great Divine Life flows everywhere.
Deut. 18: 10-12	Servant has completed his service.
Deut. 8: 13	Divination forbidden!
Deut. 8: 22	Nearly same as Jesus said, probably where he learned it. "Be ye perfect." Matt. 10:12-13.
Deut. 20: 10	When a prophet speaketh in the name of Jehovah, it cometh to pass.

Avoid the mistakes of the Hittites, Amorites and others. God's creation is a perfect creation.

170

Do not have dual thought or mixed belief. See Ezekiel, 18th chapter.	Deut. 22: 9-11
Curses. We are blinded by false beliefs. The curses were because they had forsaken the law, had not obeyed.	Deut. 24: 16
Negative side. If we do not see clearly, we get into a negative attitude. Read blessings Deut.28:1-14.	Deut. 27: 12, 15, 26
God is not hiding our clouded vision hides.	Deut. 28: 45-47
The great vision today is the Perfect Body. "He that dwelleth in the secret place of the Most High." Truth is the Omnipresence and all that that means. The secret things of the Lord are what the individual has not seen. The highest we have seen has not touched the Most High. Only what we see belongs to us it belongs to God forever, but it is ours when we see it and receive it.	Deut. 28: 58-59
	Deut. 29: 29
When we cleave to God, we cleave to Life. Moses' song, one of the Epic poems of the Bible. The Israelites thought of God as a human being. Moses was sent to Mt. Nebo. Every time we go to the Mount, high or clear consciousness, something dies --- a belief, a personal opinion. Jesus said, " Today and tomorrow I will work cures, and on the third day I will see perfection filling all." Luke 13:32.	Deut. 30: 2-3, 6, 20
	Deut. 32: 4, 10-12, 15-16, 28-30
Moses' blessing. Destroy conceptions.	

Return to the Top   Return to the Top

CHAPTER 6.

JOSHUA, THE LAST BOOK OF THE HEXATEUCH.. - p. 172

Joshua

It is the last book of the Hexateuch.

Story of conquest and partition of Canaan, containing the events of Joshua 's life after the death of Moses.

This land had been promised the Israelites, but you see there was something for them to do. We talk about standing still and seeing the salvation of the Lord, but that does not mean that there will be no activity. It means that the Lord will work in us more than ever when there is the willingness to let God work.

Although this was the "promised land," they were to put their feet on the ground before they possessed it. The feet symbolize understanding. That is what the whole tendency of this teaching is, to understand God and man, and man's relation to God. There must be co-operation with God.

For many years the children of Israel had constant battles. They were told when they went into the promised land that they must cast out every inhabitant of that land; that has seemed a cruel thing, but when we read it spiritually, it is not cruel at all, but the reverse. They were told to

172

---

kill men, women and children, that even the smallest were not to be spared. They began perfectly, and had no trouble conquering, because they had confidence in God's going forth to battle for them. However, toward the last, when there were just a few who were insignificant, they thought these did not need to be cast out, and so they were left but they were the ones who later caused them a great deal of trouble and sorrow.

The enemies we have left in the land are not merely medicines, personal beliefs, race beliefs, etc., but every critical word or unkind thought, every jealous or angry thought.

Jehovah's charge to Joshua.

Josh. 1: 3, 9, 13

Joshua, after criticising Moses for sending spies, sends two himself not a very high

plane of faith. To spy led to lies and disobedience of the Law.	Josh. 2: 1
They went from Rahab's house with lowered vision.	Josh. 2: 16
Their attitude stronger.	Josh. 2: 24
Ark --- safety in God. Israelites look upon Joshua with the same reverence they had for Moses.	Josh. 3: 13
	Josh. 4: 14
Manna was higher than the old corn. If we make our own provision, we shall be blind to the supply that is on hand.	Josh. 5: 6-12
Holy ground.	Josh. 5:13-15

173

Josh. 6: 3	Six days doing. Josh. 6:4, seven days and completion.
Josh. 6: 26	Jericho, that which stands against our new vision. Working of faith in the soul, that which appears so hard in the beginning of our new beliefs.
Josh. 7: 3, 14, 21	Taking of Ai. They began to fill with personal pride about Jericho, and said they did not need more men --- and were defeated at Ai. Sin in camp, covetousness.
Josh. 7: 25	"Stone" --- Truth. Using the Truth to destroy conception. Achan was stoned in the Valley of Achor.
Josh. 9: 3, 14, 21	Shows subtlety of ignorance; they acted on their own responsibilities.
Josh. 10: 42	Tells of five kings against the Gibeonites, history of victory.
Josh. 10: 12-14	Great soul completely illuminated by the Divine light. Israel --spiritual sense not yet conscious of perfection.
Josh. 15: 63	
Josh. 16: 10	Results of disobedience; results of the mixture.
Josh. 17: 12-18	Joseph's son's lot extended.
Josh. 18: 1	Setting up of tabernacle at Shiloh --- the Christ spirit, and the land was subdued. "Land," here, is not used in its highest meaning.
Josh. 21: 43-45	We left the children of Israel peaceful they had glimpses of quiet times. The Law always fulfills our spiritual realization, never sees an enemy

174

when we keep to the Truth. The Law sets all things right.

We are to begin to put out all fear, overbearance, wrong conceptions, anger and so on, that would say there is something besides God.   Josh. 22:

There were two and one-half tribes who did not want to cross Jordan. They helped   Heb. 3:8. conquer, and then they came back. Reubenites, Gadites, and half tribe of Manasseh. They dwelt in their own land; built an altar as witness, not for sacrifice. The great lesson of the Bible is obedience.

Joshua stricken. We cannot pin our faith to one leading thought. My highest vision may

fade and a better one take its place.	Josh. 23: 1
Not to mingle with outsiders or worldly things; not to call by name those things which we know to be unreal.	Josh. 23: 7
	Josh. 23: 12-14
Cleave unto the Lord. The way of all the earth, die.	
	Josh. 24: 12-13
"I sent the hornet before you." "Not with thy strength, but God's." 13th verse --- do not take personal pride or praise.	Josh. 24: 29-31

Joshua dies. Israel made what God had done for them the basis of their love, and we make what God is, our basis.

175

Return to the Top   Return to the Top

CHAPTER 7.

JUDGES. . - p. 176

Judges and Ruth

THE BOOK OF JUDGES.

History of the nation from Joshua to Samson. The Judges formed temporary heads in particular centres, or over particular groups of tribes; Barak in the north of Israel, Gideon in the centre, Jepthah on the east of Jordan, Samson in the extreme southwest.

	After Joshua's death, the Israelites want to know who will lead. "I have delivered the
Jud. 1: 2	land into his hand." Believe that ye have received.
Jud. 1: 19	Conception of the external contradicted in Josh. 17:18.
Deut. 20: 1	Iron charlots, unrefined metal, undevelopment.
Jud. 1: 21, 25, 27, 28	Repetition, compact with ignorance, a bribe, afflictions, to allow them to remain is for us to believe we can act contrary to what God is.
Jud. 2: 2	If we think we can find good in the external as a cause, we are making other Gods.
Jud. 2: 16-19	Help always at hand.

Jud. 3: 5-7, 9, 15   They lacked judgment, did evil in the sight of the Lord after they had seen then they turn back to their source.

Jud. 4: 1-4

Repetition. Deborah, a prophetess, the wife of

176

Lappidoth, judged Israel at that time; a powerful woman, a mother in Israel.

Song of Deborah, Epic poetry. Read. By-ways, crooked ways; high-ways, the True way.

Jud. 5:

And the land had rest forty years. Reduction of Gideon's army; the Lord does not conquer by numbers.

Jud. 5: 6

There were two types of men to consider as warriors. The manner of drinking, lapping water with tongue, and bowing down upon knees to drink.

Jud. 5:13-31

Jud. 6: 1-13

Lapping water with tongue --- water, possibilities.

Jud. 7: 2, 5, 7

Tongue --- power of our own words. Symbol -- those that speak positive words would conquer.

Jud. 8: 22-23

Gideon recognized God. They worshipped the work of their own hands. Jotham's Fable. We are not conscious of God's help while we look to other sources for help.

Jud. 8: 27

Jud. 9: 8

Apart from spiritual because he was son of a harlot. Harlot --- thought apart from Divine way.

Jud. 10:13-14

Jeptha does not represent very high development.

Jud. 11: 1

Story of Manoah and his wife, parents of Samson. The birth of Samson similar to birth of Isaac, John the Baptist and Samuel.

Jud. 11: 31

Jud. 13: 24

We usually turn to the external because it

Jud. 14: 1-3

177

Jud. 14: 5-6

pleases us. Samson turned away from Israel in choosing his wives.

Jud. 14: 14

Samson kills a lion with his hands, " Spirit of Jehovah came mightily upon him."

Jud. 15: 14, 17, 19

Samson's riddle at the marriage feast.

"And there came water thereout and when he had drunk, his spirit came again."

Jud. 16: 19, 30, 31

He trusted unwisely, his betrayal. And he judged Israel twenty years.

Jud. 17: 6

No kings, no ruling consciousness was there,

Jud. 21: 25

but a personal confusion. " Every man did that which was right in his own eyes."

Return to the Top  Return to the Top

RUTH.. - p. 178

THE BOOK OF RUTH.

Story of the ancestors of the royal family of Judah. It is considered an appendix to Judges. It shows the genealogy of King David through Ruth, a proselyte to the Jewish religion a preintimation of the admission of Gentiles into the Christian religion. Ruth was not of Israel.

I. Samuel gives the history of the Jewish church from the birth of Samuel to the death of Saul --- eighty years.

II. Samuel gives the history of David, second King of Israel, during forty years.

I. and II. Samuel are called, in the Vulgate, the I. and II. books of Kings, being two of the four books which give the history of the Kings of Israel. They are considered a key to the Psalms:

178

Return to the Top Return to the Top

CHAPTER 8.

I SAMUEL.. - p. 179

I. Samuel, II. Samuel, I. Kings, II. Kings and Chronicles

THE BOOK OF I. SAMUEL.

Call of Samuel. Answer, "Here I am" --- perfect OBEDIENCE.

1 Sam. 3: 6

Israelites believed the glory of God had departed from them when the ark was taken away by the Philistines.

1 Sam. 5:2, 3, 4

Ark --- presence of God.

1 Sam. 5: 11

Presence of God is always destruction of unbelief.

1 Sam. 7: 15

Samuel is Judge, a very clear vision for his time. The last Judge of Israel.

1 Sam. 8 5-7,
9-10, 18,
20-22

They had turned away so often they thought the kings could fight their battles for them better than God could.

1 Sam. 9: 2,
16, 17

We learn what the people did. Second verse description of Saul.

Stands for the wavering beliefs. Who shall rule over us?

1 Sam. 10: 1-4,
19-24

They ask for King after having seen the Lord as leader; this was their sin.

1 Sam. 12:
12-13,
17-22

Saul took the Amalekites.

Samuel's attitude was one of severity toward Saul.

David, the name which is given to no one in the

1 Sam. 15: 8, 9,
11, 20, 22,
24, 32, 33, 35

179

1 Sam. 16:

Bible except the great king of Israel, means "beloved of Jehovah."

There are two parallel narratives of David's life. Our first will be the Biblical account of David to be found in I. Sam. 16, to I. Kings, second chapter, and in I. Chron., 2-3 and tenth to twenty ninth chapters, inclusive.

It is the picture of David such as was present to the minds of devout Jews of the third century, B. C.

David was the youngest son of Jesse, a Judaean of Bethlehem.

We first hear of David when he was introduced to the court of Saul. The king, attacked with melancholy, " an evil spirit from Jehovah," his servants suggested that a skilful player upon the harp should be brought to soothe the king with his music.

When David came and stood before Saul, the old king loved him greatly, and the sunny-faced lad from the quiet shepherd life played soft peaceful strains upon his harp it was the music of boyish peace, the peace of freedom and the king was refreshed, was well, the "evil spirit from God" departed from him.

The two armies faced each other in the valley of Elah. Goliath of Gath, the champion, who went out from the Philistines' camp, stood and cried unto the armies of Israel, "Why are you come out

180

---

to set your battle in array? Am I not a Philistine and ye servants to Saul? Choose you a man for you and let him come down to me. I defy the armies of Israel this day. Give me a man, that we may fight together."

David, who stood by with his brother Eliab, was amazed at such daring, and in spite of his brother's reproaches, spoke to the men. "What shall be done to the man that killeth this Philistine and taketh away the reproach from Israel? Who is this uncircumcised Philistine that he should defy the armies of the living God?"

Then he bravely reassured Saul that he, his servant, would go and fight this giant, even though the king called his attention to his being but a youth and the fact of his opponent's being a giant and a man of war.

The boy told him how he had slain the lion and the bear that disturbed his flock and concluded with " this uncircumcised Philistine shall be as one of them, seeing that he has defied the armies of the living God. The Lord delivered me out of the paw of the lion and out of the paw of the bear; He will deliver me out of the hand of this Philistine."

When Saul placed his armour upon him, the young minstrel put it off. "I cannot go with these I have not proved them."

Then, as he approached Goliath; "Thou

181

---

comest to me with a sword, and with a spear, and with a shield but I come to thee in the name of the Lord of hosts, the God of the armies of Israel, whom thou hast defied. This day will the Lord deliver thee into my hand, and I will smite thee ' that all the earth may know there is a God in Israel."

.We leave the boy-hero at this point and later find him anointed king over all Israel.

We must not judge the oriental fashion in which David lived from a modern standpoint. In his day and time a man's wealth and power were to a great extent measured by the number of his wives and the size of his family.

The simplicity and tenderness of his youth ever remained with him. The bond of loving friendship between the youths, Jonathan and David, was remembered by David the King in his kindness to his friend's son Meribbaal.

When the sacred ark was brought to Jerusalem from Kirjath-jearim, David, wearing a prietly linen ephod, danced in the procession before it. His wife, Michal, ridiculed his undignified appearance; but the king, with true dignity, expressed his readiness to dance before Jehovah, who had chosen him above the house of Saul.

The prophet Nathan was a powerful factor in David's life during his reign over Israel; it was through this prophet that God's blessed message

182

---

came to David (II. Samuel, 7th chapter). The story of Bathsheba was doubtless not unknown in Jerusalem the moral sense of the people found expression through Nathan, who by means of a parable boldly rebuked David. At the death of the first son born to Bathsheba, we have in David's words of grief, "I shall go to him but he shall not return to me," the first declaration of immortality.

The story is told of Absalom's flight to the court of his grandfather, the king of Geshur. For three years he was banished, then for two years he was excluded from court, for David could show severity even to his beloved son.

Some years must have elapsed before the closing scene of David's life. We see him now in the feebleness of old age, kept within the palace, nursed by a young damsel of Shunem, named Abishag.

Nathan has to call to his memory his promise to Bathsheba that her son, Solomon, should reign after him. (I. Kings 1:21.) The old king attended at once to seeing that Solomon was placed on the royal mule by Nathan, 2adok and Benaich, and at his command conducted to the spring of Gihon and solemnly proclaimed king.

David, whose "heart was perfect with Jehovah," reigned seven years and six months in Hebron, and thirty-three years in Jerusalem.

183

With noble sentiments and a final charge to his son, the good kind David, "the man after God's own heart," died at an advanced age. He was buried in the capital, which, received from him the name of the "City of David."

After the return from the exile, the sepulcher of David was still pointed out between Siloam and the "house of the mighty men." (Neh. 3:15, 16; Acts 2:29.)

Our authorities do not enable us to say how long David continued in the position of Saul's minstrel and armour-bearer. His success in war against the Philistines, his popularity among the soldiers the love of Michal, his strong friendship with Jonathan, with whom he entered into a covenant of brotherhood, these facts are all attested by more than one passage in both narratives.

It is not easy to trace the beginning of the distrust which Saul conceived for his favorite, who had been promoted to the position of bodyguard.

The main reason for Saul's enmity is his jealousy of David's popularity and success in war, which is said to have been excited by the song of the women, who met the victorious warriors with the words, "Saul hath slain his thousands, and David his ten thousands," and the hints of a suspicion that David conspired with Jonathan to dethrone him. All we know of Jonathan goes to disprove this. Saul's towards his former favorite

184

---

increased so greatly that he purposed to put him to death. Jonathan, however, pleaded to his father David's good deeds, and on Saul's relenting he brought David out of his hiding place in the field, and presented him to his father. The reconciliation was of short duration, for, soon after Jonathan's appeal, Saul in a fit of madness, cast his spear at David as he played on the harp before him.

David fled to his home, but that night Saul sent messengers to watch the house, and, while respecting his sleeping enemy in accordance with Oriental custom, he ordered them to kill him in the morning.

David was saved by his wife Michal, who lowered him from the window. The first place visited by David in his flight is thought to be the priestly city of Nob, south of Gibeah and north of Jerusalem. To Abimelech, the head of the priests of Eli's family, he told that he was on business for the king and therefore was received with kindness. Shortly afterwards when Saul learned the details of this visit, he ordered eighty-five priests slain and the city of Nob completely destroyed. David fled from Nob to the cave or stronghold of Adullam, a place in Shepelah, west of Hebron. Here the wild country afforded him a hiding place, he was among his own tribesmen, and on the extremity

185

---

of Judah, Saul's authority was weakest. There David's outlaw days began.

David made his abode in Engedi, a tract west of the Dead Sea later he was followed there by Saul. It was here in a cave that David cut off it corner of the robe of Saul as he lay asleep; he refused to harm the "anointed of Jehovah." He followed Saul as he left the cave, and holding out the portion of his robe, showed the king how he had been at the mercy of the man whom he was relentlessly pursuing, and begged him to no longer listen to those who charged David with conspiring against him.

Saul, touched with David's generosity, pretended to acknowledge his rival's superiority, and to recognize him as the future king of Israel. But it was insincere and David's outlaw life continued for perhaps two years before he heard of the defeat of Israel and the death of Saul and his three sons.

The defeat of Israel was commemorated with mourning and fasting, while David himself expressed in a beautiful ode his grief for Saul and Jonathan. Of both he speaks in tones of warmest respect and affection his love for Jonathan is declared in a burst of passionate feeling. David could now return to his native country, so he removed to Hebron, a sacred city of Judah, accompanied by his family and his followers. The

186

Judaean elders recognized that a renowned warrior of their own tribe was more likely to defend their interests than a younger descendant of the house of Saul. It is believed that for several years the fighting continued.

David was thirty-seven years of age when the elders of the nation assembled at Hebron and anointed him King over the whole of Israel. The task imposed upon him by the election was that of freeing his country from Philistine domination.

The history of these battles, of the "mighty men" who forced their way to the gate of Bethlehem to bring David water and how the gift obtained at such a risk was too precious to drink, and David poured it out as an offering to Jehovah his many victories and final capture of Jerusalem, is a signal piece of David's genius and statesmanship as well as of his being a great warrior.

His patience with and love for Absalom during his three months' rebellion, his tact and diplomacy in the anointing of Solomon, when Adonijah was being greeted by his guests as already king, is interesting to read. We must not expect to find a saint in David, but a king, a hero and a man. No testimony to him could be more eloquent than that of the charm he exerted on all who had to do with him. Everywhere he inspired love and devotion. He accepted his misfortunes with resignation, and acknowledged them as a consequence of his sins

187

---

while he retained his trust in God's goodness. In all the varied difficulties of his eventful life, he is never without resources. He executed judgment and justice to all his people.

It was through David that the group of Israelitish tribes became a powerful nation. Israel began to feel that it had a mission in the world, and this conviction never died, even in the darkest hour. The people believed that in God's own time they would be called upon to proclaim to heathen races Jehovah's great and holy name. (Isaiah 55:3-5.)

1 Sam. 17: 47	David had no fear of anything in the external. Without proof, David trusted the indwelling power.
1 Sam. 18: 1, 2, 6, 7, 16, 20, 21, 29	Stone --- Truth. The battle is the Lord's.
1 Sam. 24:2-5	Seventh verse is a lesson not to compare. Twenty-ninth verse, Saul was afraid of David.
1 Sam. 24: 16-18	We make an enemy of things we fear.
1 Sam. 25: 1	Proved that David had Saul in his power, but was kind and did not harm him.
1 Sam. 28:	Saul appreciated his kindness for a time.
	Death of Samuel.

Saul was sore distressed and seeks among the dead, seeks the witch of Endor in the night. He asks for Samuel. Samuel said to Saul, "Why hast thou disquieted me, to bring me up"? He tells Saul what is coming to pass.

188

Mediumship on very low plane of development. Compare Luke 29:30-31.

Isa. 8: 19-20

Death of Saul. Philistines slew his sons and his armour-bearer kill themselves.

1 Sam. 31:1-6

Return to the Top  Return to the Top

II SAMUEL.. - p. 189

THE BOOK OF II. SAMUEL.

The books of Samuel, Kings and Chronicles relate to the same histories, and often one explains what is not plain in another.

I. Chronicles contains an abstract of the history of the time it was written. It is an abridgment of all sacred history, especially from the origin of the Jewish Nation to the return from the first captivity.

II. Chronicles contains historical particulars not mentioned elsewhere.

I. and II. Chronicles are reviews of history of the kings of Judah and Israel.

The first king of Israel slew himself. With each of us there is some king ruling our thought, though we may not want it to. Something rules our thought or we should never show a sign of imperfection. Whatever it is, it will slay itself in time, but not until we are ready to give it up. It helps us to read the Bible, and we shall not be afraid to read that God changed his mind or that God's chosen people did wrong (those that thought they were chosen), when we know how the voice of God has been interpreted according to the consciousness of the interpreter.

God did not want David to build a temple. David said there is just one thing he desired, " To dwell in the house of the Lord forever." The house of the Lord is the presence of God; to dwell

190

in the house of the Lord is to dwell in the consciousness of God's presence.

Solomon was given a choice of blessings. He chose "an understanding heart." God was pleased, and said, since you have not asked long life or riches, behold, I will give you an understanding heart. Solomon had wisdom, but not love. Jesus had wisdom and love combined, and that was why he was a perfect manifestation.

Return to the Top   Return to the Top

Help us process more books to support your journey

Click here to give your support to New Thought Library

THE BOOK OF I. KINGS.

This book contains the history of the nation from the death of David and Solomon's accession to the destruction of the kingdom of Judah. I. and II. Books of Kings form one book in Hebrew manuscript; they include the whole time of the Israelitish monarchy, exclusive of the reign of Saul and David.

David's favorite son was Solomon. The great king was so impersonal that he rejoices in his son's being made king.   1 Kings 1:34

Death; David slept with his fathers.   1 Kings 2: 10-12

Solomon made affinity with externals (Egypt). Wisdom, unguided by love, will go out to   1 Kings 3: 1, 3, 4
strange women. Results in bondage always.

To come above all the hills of difficulty is to come up in the height of consciousness.

7

191

1 Kings 3:16   The true mother consciousness does not want duality.

Divine Science Bible Text Book by A. B. Fay - Read the Complete Text for free at NewThoughtLibrary.com

1 Kings 4: 30-34	Solomon's wisdom.
	Doing this for his own personal glory.
1 Kings 5:5	Seven years building the temple.
1 Kings 6:38	Thirteen years building his own home, house for Pharaoh's daughter.
1 Kings 7:1	Tells of dedication of temple. 17th, 18th and 19th verses. We often hear God speak when we
1 Kings 8:	
1 Kings 9:1	want our own way. 27th verse, highest vision Solomon had --- he began to see God as universal.
1 Kings 10:1	He heard God according to his desire.
1 Kings 11:1-3	Queen of Sheba visits the king to prove him with hard questions. She is deeply
Kings 11:28, 31,32,33,35, 36,41,43	impressed. Verses 7, 10, 21, 23. Vast wealth of Solomon. Wisdom which God had put in his heart.
1 Kings 12: 1, 4,6, 20, 21,	Personality secret of Solomon's downfall; he had to grow through the process of the "consuming fire" because he turned to externals.
	Jeroboam, Ahijah the prophet. Solomon's enemies.
1 Kings 13: 4-6	Jerusalem, heart of love and goodness. Light of Divine Nature. Death of Solomon.
1 Kings 15: 23-24	Rehoboam, son of Solomon, follows evil counsel. Israel rebels and Jeroboam is made king.

Withered hand, Divine Healing, instantaneous healing. This is quite equal to the healing of the withered hand by Jesus.

Asa was a good king. "And Asa (diseased in

192

his feet) slept with his fathers." (There are three kinds of sleep: the natural sleep at night, the sleep which is called death, and the sleep of the unawakened consciousness. Our souls are our physicians, and the world is coming to know it.)

Asa, in his disease, sought not to the Lord, but to the physicians; result, "And Asa slept with his fathers."	2 Chron. 16: 11-13
Elijah fed by ravens; birds --- soaring thoughts. Divine supply under all circumstances if we know how to trust.	1 Kings 17: 5-6
Elijah raises the widow's son from the dead. 24th verse, the same was said of Jesus. John 3:2, John 16:30.	1 Kings 17: 10-16
	1 Kings 17: 17-24
"How long halt ye between two opinions? If the Lord be God, follow him; but if Baal, then follow him." Water --- divine possibilities. Love --- consuming fire.	1 Kings 18: 20-40
Jesus also reached point of unfoldment in forty days' fast.	1 Kings 19:8
Elijah is comforted by angel; casts his mantle upon Elisha. Mantle --- covering of the spirit which unfolded.	1 Kings 19: 16-21

Return to the Top   Return to the Top

THE BOOK OF II. KINGS.

Chronicles of the Kings of Israel.

2 Kings 1:18

"And it came to pass when they were gone over that Elijah said unto Elisha, 'Ask what

I   2 Kings 2: 9-12

193

---

shall do for thee, before I be taken away from thee.'"

And Elisha said, "I pray thee, let a double portion of thy spirit be upon me."

Elisha's answer was, "If thou see me when I am taken from thee, it shall be so unto thee; but if not, it shall not be so."

Matt. 20:21-23	Elijah saw the law; if Elisha had the open vision, the spiritual understanding of Elijah's transformation, it would prove that he was one with Elijah in realization, and his desire would be fulfilled. Jesus gave practically the same teaching to the mother who asked that her sons might sit next him in the kingdom. Paul gave the same lesson in Rom. 8:11.
2 Kings 6: 15-17	
Ps. 68: 17	
2 Kings 13: 14	And it came to pass as Elijah and Elisha talked,, that, behold, there appeared a chariot of fire, and horses of fire, and parted them both asunder and Elijah went up by a whirlwind into heaven. And Elisha saw it!
2 Kings 2:23-24	
2 Kings 4: 1-6	The writers of the Bible express their deepest experiences in symbols. Chariots and horses denote the Presence of Omnipotence.
2 Kings 4:35	Rebuke given (cursed --- rejected) to those not on so high a plane by your higher understanding.
2 Kings 4:42	

Elisha and widow's oil. Set your vessels --- do it for yourselves.

Divine Healing.

Multiplying food. Also Mark 6;37-44,

194

Healing. "He shall know that there is a prophet in Israel." Naaman's leprosy healed.15th verse --- voice within said, "Humble thyself" personality. Since Divine Healing was freely practiced during the ages before Jesus, we see that this power did not belong exclusively to Jesus, neither did it originate with Jesus, nor was it a special gift to him, whereby he might prove himself divine.

2 Kings 5:3, 8-14-15

Jno. 14: 12

Matt. 21: 21

Jesus taught that this power belonged to all and to every age, if there were faith. The strongest proof of this is the fact that today all who believe in the Omnipresence of God, are doing these works of faith in the raising of men from the dead sense of sin and disease.

2 Kings 6: 15-16-17

2 Kings 6: 28-29

History of Israel to destruction of the temple. Prayer. The only prayer is our opening our vision to the Truth of things. Seeing is believing; believing is accepting.

Famine in Samaria. Jerusalem, capital of Judaea; Samaria, capital of Israel. Israel and Judah --- two tribes. Great wars between these two tribes. Kings were said to do wickedly.

2 Kings 18: 3-4, 9-12, 21

2 Kings 19:2

Good kings always compared to David, he was so impersonal.

Hezekiah, King of Judah. (Chapter same as Isaiah 37.) First mention of Isaiah.

195

---

2 Kings 20:5	(Parts of this chapter same as Isaiah 38.) "I have heard thy prayer, I have seen thy tears; behold, I will heal thee." Thus spake the Lord to Hezekiah. Yet Hezekiah demanded a sign. (II. Kings, 20:8.) In faith, this was equal to Isaiah's addition to God's promise, verse 7. Isaiah, not God, said this.
2. Kings 23:21-26	
2 Kings 23: 32-36	Jehovah strengthened.
2 Kings 24:	To do evil is to believe in evil.
2 Kings 25:	Beginning of the destruction of Jerusalem by Nebuchadnezzar.

(Chapter same as Jeremiah 52.) Demolishment of those who had fallen away from the Lord.

Return to the Top  Return to the Top

I & II CHRONICLES.. - p. 196

THE BOOK OF CHRONICLES.

First and Second Books of Chronicles, the record made by the historiographers of the kingdoms of Judah and Israel. They are the official history of these kingdoms.

God's promise to David. David's final charge to Solomon and to the people.

1 Chron. 1:1;
10: 13-14. 17:
10-15

1 Chron. 29:10-
20

196

Return to the Top   Return to the Top

CHAPTER 9.

EZRA.. - p. 197

Ezra, Nehemiah and Esther

THE BOOK OF EZRA.

Story of the return of the Jews from the Babylonish captivity and of the rebuilding of the temple. Ezra, son of Seraiah, born in Babylon.

Was called a ready scribe of the law of Moses.

	Ezra 7:1-6
Wall of Jerusalem stands for the very heart of things in spiritual sense; a spiritual city,	Prov. 24: 30-31-32
broken walls very suggestive. Why, then, was a breach in the wall?	
When Israel sinned she was given to her enemies. When we err in ignorance, in our decisions, the law gives us to our enemies, our mistakes.	Ezra 1: 1-4
	Ezra 2: 4-6
If there is any breach in our walls, the one who repairs it must be of pure nature.	Ezra 3: 8-10-12
Cyrus' proclamation.	

Children that go up out of captivity over 50,000.   Ezra 5:12

Rebuilding of the house of the Lord. Artaxerxes stops the building of the temple. Darius,   Ezra 6:
king of Persia, told them to go on. He tried to see why the building had been stopped.

Kings of Persia are recognizing the God of Israel.   Ezra 7:
                                                                                         11-13, 15

197

Ezra 9:6-13-14   Ezra works a wonderful reformation.

Ezra 10: 6, 7,9,   First step in reformation, to bring our thoughts together to the great center of Truth.
18, 44

Return to the Top   Return to the Top

Help us process more books to support your journey

Click here to give your support to New Thought Library

NEHEMIAH . - p. 198

THE BOOK OF NEHEMIAH.

Nehemiah was cup bearer to Artaxerxes. Nehemiah, repairer of the wall in Jerusalem. Omnipresence is our
stone wall and breaks are where we do not see God's presence.

Neh. 1: 11   Prayer concerning Jews left at Jerusalem.

Neh. 2: 5-6, 17   Gets permission to visit them.

Neh. 6:15-17   The wall is finished.

Neh. 8: 1-3   Ezra reads the Law to the people.

Neh. 9: 27   Saviours.

Neh. 10   Covenant to keep the law and maintain the temple worship.

Neh. 12: 27   Dedication of the wall of Jerusalem.

Neh. 13: 3   Gave up dual beliefs. Captivity not in place, but in mentality.

Return to the Top   Return to the Top

THE BOOK OF ESTHER.

Poetical books, including Job to Song of Solomon, were written at various times, some being of earlier, others of later date than the historical books. They are classed together, partly because they are in Hebrew verse, but mainly because they formed the devotional books of the Jewish church.

198

Story of Jewess who became Queen of Persia and saved the Jewish people from destruction.

Esther contains episode in history of those Israelites who did not return from captivity, shows their moral decline.

Although God's love never forsakes his people and he delivers them, His name is secret among them.

Incident supposed to have its historical position between sixth and seventh chapters of Ezra. Ahasuerus conjectured to have been Artaxerxes. Impossible to identify Esther with any queen mentioned in history; probably a favorite concubine to whom the title was accorded. Author of book unknown; probably Mordecai. No one else was in possession of knowledge of details. Attributed to Ezra, who may have brought it from Babylon to Jerusalem and added it to the Canon, Written in Hebrew, though additions added in Greek. The feast of Purim remains to this day as an evidence of the truth of the story and the book has been esteemed Canonical by Jews and Christians.

199

Return to the Top Return to the Top

CHAPTER 10.

Job, Psalms, Proverbs, Ecclesiastes and the Song of Solomon

THE BOOK OF JOB.

There is a doubt as to when Job lived, probably about the seventh or fourth century B. C. It is an allegory and there is an application to be taken from it for ourselves, an inner meaning which will help us.

Job 1: 6   We are all sons of God, and we come to present ourselves before, God. Each soul can say, "I am the son of God", but can I go before God with a perfectly unmixed consciousness? I think very few of us do this.

Personality is called Satan sometimes; any claim of separation from God can be called Satan.

Think of the omnipresence of God, let that be the basis of your decision; then ask yourself if there is a spot in the universe where you are willing to say God, is not! Sometimes we think of Satan as unbelief. (Belief and unbelief are simply conditions of our own mentality. To the extent that one expresses anything lower than perfection there is unbelief; that would account for everything

200

---

that comes to us. It was thought that Satan brought all these things upon Job.

Jesus sent the seventy out and they came back, saying that even the devils were subject to them and were rejoicing in what. they were able to do in the Divine name, and Jesus said to them, not to rejoice in that, but in that their names were written in heaven. That means they were sons of God.

We get the fruit of our own belief because we act upon our belief. Rise out of your sense of limitation, such as pride, grief, sickness, and so on. To be well permanently, we must have a basis for our health. Open your eyes to see the harmony of God's Presence. We all have the Divine inherencies or qualities that belong to God. Let us not seek praise, it is personal. It is the individuality we are bearing witness to as the expression of God. The body of God's perfection is here in me. Acknowledge God's wholeness. You do not possess a thing consciously until you claim it!

Job shows the development of a soul from the worshiping of an outside or personal God to the finding of the God within.

Three friends were on the same plane of understanding as Job, but he unfolds away from them. Three here, stands for collection of beliefs. As long as men have whirlwinds within themselves (strife) there will be whirlwinds without. Satan,

201

---

human personality claiming something for itself apart from God. John 8:44.

Job 1:1-5 ---

1st: Fear of God, offered sacrifices.

2nd: Belief that God is changeable.

3rd: Did not stand, but sat (2:8) in the ashes.

4th: Duality, belief in evil.

5th: Fear, personal.

Fear is the beginning of wisdom. Job believed this story of himself, it was a part of his development, as a type.

Fourth verse means his light, this day, beginning of light. Each individual knows he has these two attitudes, fear and good, an argument between God and Satan. Satan, adversary, unbelief. The highest never argues, pure consciousness knows.

Job 1: 1

Job 1: 21

Job 's opinion of himself. He based his perfection on what he had and what he had done, but the real basis of Perfection is in God. What I Am is the basis. Job listened to the argument. If everything is swept away, it is according to Law, and after listening he had the experiences resulting from the argument when we waver in thinking, we always have unhappy experiences.

"The Lord gave and the Lord hath taken away blessed be the name of the Lord." This statement has become a fixed formula in many systems

202

---

of religion. We do not believe the Lord ever takes away anything. Yet was it not better for Job to say that than to complain of God I In a degree that was good, but later he says, "I have uttered things which I understood not". That is one of them.

The Lord is changeless and does not take away. God is always giving, but the turning is on our side, our attitude, the soul's attitude toward God.

When our eyes are opened to unity, our loved ones will not go beyond our sight.

Satan touches Job's body. He has been very complaisant under the loss of family and cattle, but Satan says, "All that a man hath will he give for his life, but touch his bones and his flesh and he will curse thee to thy face".

Job 2: 4-5

Job 2: 7

Why have we always been taught that God sends sickness and that we must be patient under it? If God sends sickness for a purpose, and we are to be patient under it, why do we try to get rid of it

"So Satan went forth from the presence of the Lord, and smote Job with sores and boils." Job forgot the presence of the Lord; Satan could not have afflicted Job with sores in the presence of the Lord; in the consciousness of the presence of the Lord. Of course, being away from the presence of the Lord is only in belief.

203

Job 2: 8

When we find our bodies stricken, lift up thought, take the positive stand and do not waver. Job sat in ashes; gave up.

Job 2: 9

Then his wife said, "Dost thou still hold fast thine integrity? Renounce God, and die".

Job 2: 10

Job answered, "Thou speakest as one of the foolish women speaketh. Shall we receive good at the hand of God and shall we not receive evil?" Job did not rebuke God, but saw two powers, good and evil, duality.

Job's three friends came to him. This was the condition of his own mentality, nothing from the external; no friend can come to us whom our own mentality does not bring. It is what we are giving out that comes back; our first work is with ourselves.

Now these three friends came and sat in silence seven days. Seven means a perfect period. After seven days of silence, Job cursed the day he was born. Then Eliphaz said, "The judgments of God fall on the wicked," and tells Job not to despise the chastisement of the Almighty; that he who sins is punished; he urges Job to accept his punishment and value it. "Whom the Lord: loveth, he chasteneth." What is chastening? It is purifying. Should we despise the consuming fire? Read Job 5:19-27.

Bildad shows how God punishes the wicked and preserves the righteous; that is the consuming

204

---

fire. Bildad had Job's punishment in mind, but he spoke a great truth. The destruction of everything that hurts or harms is the working of the law.

Job acknowledges the power of God, but says,"My soul is weary of life." He is rising in consciousness, he is speaking to God, telling him to let him alone, to let him have a little comfort.

Outside friends mean all the thoughts that believed in evil came up.

	Job 2:11
Sank to his level, sat down with him.	Job 2:13
A day, breaking of light; cursed his day, rejected the old light and was willing to give up what he knew, old beliefs, a turning point.	Job 3:1
Words of three friends, condemnation and rebuke, condemnatory thoughts within Job. Do not condemn yourself.	Job 4: 3-9
	Job 5:17
These spoke according to their light. Job answers, beginning of his acknowledgment of erring.	Job 6: 24
God is dealing justly with Job, but Job gradually changes his attitude, which is the cause of his trouble. Begins to see that it is not profitable to know so much that is not true. Job did not need condemnatory thoughts the friends did not tell him to rise.	Job 7: 16
	Job 8:
Job confesses God's power, and expresses fear again.	Job 9: 4-28
Maintains his integrity.	Job 10:

205

Job 11:	Zophar reproves Job for justifying himself, "Canst thou find out the Almighty unto perfection?"
Job 11: 17	Duality shines all through this, these men are mixture of Truth and, misconceptions.
Job 12: 2-3	Job gets sarcastic, "No doubt but ye are the people, and wisdom shall die with you.
Job 12:10-13	But I have understanding as well as you; I am not inferior to you: yea, who knoweth not such things as these?"
Job 13: 3-5	Praises God for what he is and what he does.
Job 13: 15	Job silences his beliefs or voices that condemn; rises to greater faith in God.
Job 13: 23	We are trying to know God, trust, think of life, not death.
Job 14: 1	Make me to know my transgressions and my sins, not bemoan it, but lift it up by
Job 14: 16	understanding. To count up is a low plane of development. Try to see the other, all in

	God.
Job 15 6,8, 10	"Man that is born of woman," not of God, shows unbelief. Man is born of God.
Job 16: 2-5	God takes care of sin.
Job 16: 12	Eliphaz, aged men stand for old beliefs. Job is speaking on a higher plane.

Job reproves the three friends.

At ease, Job finds out he is "at ease," though not from a basis of understanding.

206

Job rises to a clear realization, he catches a glimpse of his real self, his individuality.	Job 16: 19
Job appeals from men to God. He sees that God does not send affliction. "Light of wicked put out," the claim that personality (Satan) makes.	Job 17:
"Satan shall be driven from light into darkness they that come after shall be astonished at his day, as they that went before were affrighted."	Job 17:13-16
	Job 18: 5-6
Astonished at the power given the claim of personality when the consciousness of individuality has come.	Job 18: 18-21
Complains of cruelty, but Job begins to lay down his personality and to see that he cannot claim anything as Job, but of God.	Job 19: 2, 6-9
	Job 19: 14, 20-21
Beginning to hear the call that he must drop personal beliefs his appeal to the pity of his friends even while protesting against their reproaches.	Job 19: 25-26-27

A wonderful conviction of Truth. "I know my Redeemer liveth, and he shall stand at the latter day (later consciousness of Truth) upon the earth." Even his body is lifted up. "Yet in my flesh shall I see God, whom I shall see for myself my reins be consumed with earnest desire for that day" --- the light of understanding. "Latter day," later consciousness of Truth. "All flesh shall know God," shall be known as divine.

207

Job 19: 28-29	"Root of the matter is found in me." Job is seeing that the conditions are the result of his own belief. "Be ye afraid of the sword." The sword of Truth. Matt. 10:34. Job is reasoning from the surface conditions of life; sees for himself the end of personality in the grave. It is its own destroyer. "God is no respecter of persons," each soul is born of God but the unfoldment into the Truth of birthright is attended by many man-made laws, beliefs and opinions. Ecc. 7:29
Job 21: 3, 23-27	
Isa. 14:11.	
Job 21: 3-30	All wickedness (ignorance) shall be destroyed. 1 Cor. 3:15.
Job 22:21-25, 28, 30	Eliphaz's third speech is a powerful appeal of the spirit to Job for realization of God's power and love; even to his own individual expression of his conscious unity with God. "Thou shalt also decree a thing, and it shall be established unto thee; and the light (or understanding) shall shine upon thy ways."
Job 22:21-23, 26, 29	

"Acquaint now thyself with Him (God) and be at peace --- receive the law from His mouth and lay up His words in thy heart; thou shalt put away iniquity (ignorance) far from thy tabernacles; then shalt thou lay up gold as dust --- the Almighty shall be thy defense and thou shalt have plenty of silver." Deut. 8:18. "For thou shalt have thy delight in the Almighty," in the consciousness of the Omnipresence.

208

"When men are cast down," thou shalt have compassion, lift them up, even the "humble person," the one who has the least realization of the truth of their relation to God, their Father.

"Deliver the island of the innocent"; by the purity of "their hands" wilt thou take by the hand those who believe they are separated (island) from their Source, God, and assure them of the truth of Omnipresence.

Job's earnest desire to know God.

Bildad, personality still claiming a place, declares that man cannot be justified before God.
Job 23: 2-6, 10

Man is not born of woman, man is born of God.
Job 25:

"Worm of dust" attitude is being pushed into the background of ignorance, whence it came. The realization of absolute identity with God, our Father, could mean only a deep and abiding happiness.
Job 25: 4

Job 25: 6

Shows God, the Infinite, bindeth up waters in thick clouds.
Job 26: 8-10

Job holds fast to righteousness, right thinking. Man's skill in finding treasures of the earth.
Job 27: 6

Wisdom a greater treasure, whose source is God. Job had not yet risen to the heights. His friends cease to answer. Passing of Job's thoughts out of condemnation.
Job 28: 1-6

Job 31: 40

Elihu was ready to put out words of condemnation. They cannot give Job a ray of light.
Job 32: 1

209

Job 32: 6-10

Elihu represents the new recognition of Spirit; the new voice, the new vision.

"Multitude of years should teach wisdom -- but there is a spirit in man and the inspiration of the Almighty giveth them understanding --- great men are not always wise; neither do the aged understand judgment." Truth is uncovered by understanding. He who turns to the Secret Place in his own divine nature for all knowledge will walk with God. "All thy children shall be taught of God." Isa. 54:13; Jer. 31:33-34.

Job 33: 1-5

Elihu showing the higher things in order to stand up.

Job 34: 11

Man receiving according to his own attitude. Our own way of looking at Truth is our conception. If we did not have Truth, we could not have conceptions.

Job 34: 15

Job 35:

All belief in "flesh" as something apart from Spirit will perish and man shall turn again to his Source, the one eternal substance. The Presence, the Spirit of God, is to speak to Job.

Job 36: 4

Job 38: 1, 3-8, 36

"He that is perfect in knowledge is with thee." The call is, get up out of the ashes. God sets forth Job's ignorance of creation, constitution of earth, light, and so on. First words of God that Job heard. Job answereth the Lord. Jehovah condemns Job's three friends.

210

Job 40: 6-7

Job 42:2-5, 7

Job prayed for his friends, redemption of condemnatory thoughts. We must redeem every thought. The Law of God found Job when he dropped the personal attitude.
Job 42: 10

Return to the Top   Return to the Top

Help us process more books to support your journey

Click here to give your support to New Thought Library

THE BOOK OF PSALMS.

Hymns to God and precepts for human life.

The Psalms are divided into five books.

First division --- Book I, 1-41 chapters, are personal; those of the second division, Books II and III, 42-89 chapters, are national; those of the third division, Books IV and V, 90-150 chapters, liturgical.

The predominating tone and spirit of the book is religious. It is concerned with the deep elemental ideas of religion; God, man, the relationship of man to God, the human soul in all of its stages of unfoldment, and the Divine power and grace in all of its aspects.

The ethical view of the Psalmist was limited. He was the child of his own age; yet the wise and devout man knows that in the Psalms as nowhere else, except in the words of Jesus, one is brought directly into the presence of the Living God, a God that is lofty and pure, gracious and tender.

The Hebrew Psalmist might seem to be a child by the side of the Hindu sage and the Greek philosopher, but neither of these could sound the

211

---

human heart as he has done. The littleness and greatness of man are there discerned, almost unconsciously to himself, by the poet, because his eye was fixed, not on man, but on God.

Ps. 7: 11
"God is angry with the wicked every day." The "wicked" are those who do not realize God's presence and power and in their ignorance disobey the law and feel the result.

Truth perceived is ever destroying these false impressions. Expression, from within. Impression, from without. Sin is ignorance there is no escaping from the law of punishment, except by knowledge.

Ps. 23:

Gen. 49: 24
"Verily, verily I say unto you, I am the door of the sheep. By me, if any man enter in, he shall go in and out and find pasture. I am come that they might have life and that they might have it more abundantly. There shall be one fold and one shepherd." Jesus' words, a realization of the deep meaning of this psalm.

Gen. 46: 34
"Shepherd the stone foundation of Israel."

"I am the good Shepherd."

"Every shepherd is an abomination to the Egyptians." A fulfillment of the Twenty-third Psalm in the words of

Jesus.

The Lord is my Shepherd; I shall not want."

"I am the Good Shepherd."

I shall not want rest: "He maketh me to lie down in green pastures."

212

"Come unto me all ye that labor and are heavy laden, and I will give you rest."

I shall not want drink: "He leadeth me beside the still waters."

"If any man thirst, let him come unto me and drink."

I shall not want forgiveness: "He restoreth my soul."

"The Son of man hath power on earth to forgive sins."

I shall not want guidance: "He leadeth me in the paths of righteousness for his name's sake."

"I am the way, and the truth, and the life. Follow me."

I shall not want companionship: "Yea, though I walk through the valley of the shadow of death, I will fear no evil; for thou art with me."

"Lo., I am with you always, even unto the end of the world."

I shall not want comfort: "Thy rod and thy staff they comfort me."

"I will not leave you comfortless. Ye are in me, and I in you."

I shall not want food: "Thou preparest a table before me in the presence of mine enemies." "I am the bread of life; he that cometh to me shall not hunger."

213

I shall not want joy: "Thou hast anointed my head with oil." " That my joy may be in you, and that your joy may be filled full."

I shall not want anything: "My cup runneth over."

"If ye shall ask any thing of the Father in my name, He will give it to you."

I shall not want anything in this life: "Surely goodness and mercy shall follow me all the days of my life."

"Seek ye first his kingdom and his righteousness, and all these things shall be added unto you."

I shall not want anything in eternity: "For I will dwell in the house of the Lord forever."

"Whither I go thou shalt follow * * * that where I am, there ye may be also."

Ps. 72: 20

1 Sam. 2: 1

Hab. 3: 1

Prayer here is understood in its broadest sense, any turning of the heart towards God in petition or in praise is called prayer. Duplicate editions of same Psalm --- Ps. 14:53; Ps. 40:13, 17; Ps. 70:108 Ps. 57:7-11; Ps. 60:5-12.

Editorial note in Ps. 72:20. The prayers of David, the son of Jesse, are ended, seems to prove conclusively that the compiler of the collection in question knew of no other

Ps. 17: 86, 90, 102, 142

Davidic Psalms.

In the 91st Psalm we read that he that dwelleth in the secret place of the Most High shall

214

---

abide under the shadow of the Almighty. We can abide in the consciousness of the Presence of God.

"For thou didst form my inward parts: Thou didst cover me in my mother's womb. I will give thanks unto thee, for I am fearfully and wonderfully made:

Ps. 139: 13-16

Wonderful are thy works; and that my soul knoweth right well.

13th

My frame was not hidden from thee, when I was made in secret, and curiously wrought in the lowest parts of the earth.

14th

Thine eyes did see mine unformed, substance; and in thy book they were all written, even the days that were ordained for me, when as yet there were none of them."

15th

16th

--- American Standard Version.

Ps. 139: 13-16

"For thou hast possessed my reins: thou hast covered me in my mother's womb. I will praise thee; for I am fearfully and wonderfully made; marvelous are thy works: and that my soul knoweth right well.

13th

14th

My substance was not hid from thee, when I was made in secret, and curiously wrought in the lowest parts of the earth.

15th

16th

Thine eyes did see my substance, yet being unperfect; and in thy book all my members were written, which in continuance were fashioned, when as yet there was none of them."

--- King James' Version.

215

No sacred book of any nation has solved the fundamental problem of all religions, the true relationship of man to God, the Source and Cause, as has the Psalms.

Rom. 8: 12

Ecc. 11: 5

The freedom and power of Divine Science is a shining light penetrating and warming the entire world with the "life more abundant," which Life is ever enjoyed with the consciousness of Being.

Eph. 2: 8

The entire religious world is undergoing a mental change, preparatory to perfect adaptation to true Theology, the science of God in creation. In Divine Science is taught the relationship that creatures hold to the Creator, that living forms hold to Supreme Being.

Truth is now leavening the whole lump and is revealing anew the absolute idea of the Christ and of primitive Christianity in its original God-given glory and power of healing.

No expression can be superior in quality or greater in quantity than the Expressor. No effect can be higher in its nature than its source and cause; this is a basic principle to be adhered to in the study of Truth, hence the principle of equality lies at the bottom of this lesson.

The basic truth of the declaration by St. John in his account of creation is that the word that is with God and is God becomes flesh and dwells among us. As the now can never cease, but continues for all time, we know that what has been, is now, and what is now will continue to be.

The first thing to grasp is that the I Am is Spirit or Mind, the Expressor of bodily existence; and that nothing can be expressed that is not before it is manifested.

Thus it is a self-evident truth that I, the individual, am one with God, the Father.

The record of the word that is with God and is God, becoming flesh, testifies of self-expression; Infinite Mind revealing itself unto itself; so in the true Christ incarnation, we are clothed with Spirit, not unclothed, but clothed upon in the same sense that Jesus referred to when he said, "That which is begotten of Spirit is Spirit"; creation is expression. There is no overcoming but that of coming over from theories unto the truth of God, the only creative cause, when it is said, "What is made is Mine."

"I and my Father are one." "All that thou hast given Me will come to Me." "All thine are Mine," in at-one-ment. Verily, "When that which is perfect is come, that which is in part is done

John 10: 31

John 17: 10

1 Cor. 13: 10

away." There is nothing Godly but that which is like God; there is nothing ungodly but that which is unlike God; and the Creator does not begin His work with the view of its becoming more perfect through experience than He has made it; it is ever intended, therefore, that man should

rely upon his innate power and possibility, his present perfection of Life, Substance, Intelligence and Power. That which is perfect is not always complete, while that which is complete must be perfect. A child is perfect but incomplete. A seed is perfect; in its unfoldment it is completed.

The "beginning" spoken of in all the Spiritual records of creation is creative action it is the expressing and speaking forth of what the I Am is.

In Divine Science we learn what Selfhood is and where it is based and that we are free from worldly theories only in Truth. We see evolution as a mental process and interpret that process from the basis of God as Source and Cause.

Evolution consists of a theory of the evolution of form, or of reproduction, the higher forms evolving from the lower. The variety of form, color and character is produced by what is called natural selection. Nature's choice, that the origin of species is caused by circumstances (not a fixed law), that the intermixing of certain types, or forms, produces the different species, hence the variety of form. This theory of evolution is one based wholly in form and observation. It is asserted to commence with the simplest known form, every succeeding form is due to and is the result of the preceding form; that is, forms make forms, or the source of form is form, that the more complex

has evolved from the more simple one; this theory claims form to be both cause and effect, the cause being inferior to the effect, hence the result is considered to be superior to its cause. Yet it can be truthfully said that no law is known by which an effect or expression can be greater than its source and cause.

Another theory set forth differs from this one only in that it is a more interior process; namely, what is termed the evolution of the soul.

The starting point is in cosmic light, primal expression of Spirit. The beginning of the evolution of soul commences with an expression of Infinite Spirit, called cosmic light. It is to be noted here that the reasoning begins in the expression or effect instead of the Expressor or Cause; this effect, cosmic light, finally evolves the soul of a planet, then produces from this soul in regular order, the souls of all things manifested upon the planet; beginning with the mineral, then vegetable, then animal, then the lower to the highest consciousness in the human form, the human soul.

This theory has been largely considered by many who are striving "to rend the veil of sense and pass into the realization of things which lie above and beyond this shadow land."

The origin of man, the true conception and

219

birth of his body, is not found in any of these theories.

The two illustrations given are essentially the same in that both deal with the evolution of form, the form of the one being physical and visible to the natural eye, and the other form being psychical and invisible to the natural eye. One calls form, body; the other calls form, soul. Both claim that the lower form is on its way to the higher, that the higher form is always the outgrowth of the lower, hence, in both theories, effect is presented as being superior to cause. Yet to all true reasoning, the adage is true, that a stream cannot rise higher than its source; that the expression of a source and cause can never be superior in quality or greater in quantity than that which has given it expression, that the unfolded can never become superior to the unfolder, or overcome and set at naught its source and cause. Jesus said, "That which is born of flesh is flesh."

All theories having form as a source and cause by which to account for creation profiteth nothing; they constitute the flesh that profiteth nothing, of which it is said: "If ye live after the flesh ye shall die, but if ye, through the Spirit, do mortify the deeds of the body, ye shall live."

We are not debtors to the body that we should make it the source and cause of being; of our individuality

220

and character; of immortality and spirit, of consciousness.

Paul's advice to the Ephesians is good for us to follow, i. e., put off the former conversation, the old man, which is corrupt, according to the deceit of sense. If we do this we shall set aside all theories based in a supposed physical causation; false conclusions are what Paul refers to as deeds of the body that we are to mortify, for be speaks of false conversation which is no more than erroneous belief. This is to be done that we may prove that "By grace we are saved, and that not of ourselves, it is the gift of God."

Hence, to live after the Spirit here and now is to recognize and know that there is no other causation. That is what Paul referred to in the first chapter of Romans; "When they knew God, they glorified not as God, neither were thankful, but became vain in their imaginations --- who changed the truth of God into a lie and worshiped and served the creature rather than the creator."

This worshiping and serving the creature is believing that our forms are the product of lower forms or of other forms, or that creatures are the source and cause of creatures. Let us conclude that all these theories and conditions that are born of flesh are idolatrous and profit nothing. We reason from a basis; we argue from appearances.

221

Genesis --- Verily, we need to revive and to realize each for himself, the grand old record of the Genesis of creation, and come, every man and every woman of us, to the Edenic order and law of our Being. It is folly to suppose that the lines of investigation and discovery of Truth end with a knowledge of construction of form.

The individual refuses to rest content with the claim that the substance and reality underlying both ideal and visible (phenomena) is unknowable. The mere recognition of "an infinite and eternal energy, whence all things proceed," in and of itself compels men to try to find out something of the nature of that energy. Consequently, while not discarding the old records, the spiritual statements made of God manifest in the world, but working in direct harmony along the lines whence the fathers gathered material for their conclusions, we would be led to ask, what will come of it? We cannot be dogmatic in our answer as to the outcome, but will say that every new discovery tends to affirm both the oldest and the newest statements of unity.

If, in the language of natural science, there is "an Infinite and Eternal Energy whence all things proceed," the record is true, though stated in different language, that in the beginning was the Word and the Word was with God, and was God,

222

and without it was not anything made that was made.

Also, the one Eternal Energy from which all things proceed sustains the statement that as it was it is now and evermore shall be and that in God we live, are moved and are ever being.

The oneness of substance also sustains the record of the Genesis of creation, that there is only God and His Word, that the Fatherhood of God includes the brotherhood of man.

The consciousness of one Source does away with all dual doctrine, hence with all dual results and sense limitation.

That which fills all space is Spirit, hence all things are found to be in a state of at-one-ment with their Source and with each other. Spirit, therefore, carries within its infinitude of love the whole universe as Itself expressed.

Of Truth is born: Its expression and manifestation which is Truth also.

Of the Source is born: Its activity and result --- like the Source.

Of Divine Mind is born: Its thought and word, all perfect.

Of Spirit is born: Its living soul and body, all Spirit --- for like begets like.

To know the truth of our existence is to know that it is the living word; that our whole body is life and that we can say, "Whatever flesh we

8

223

bear, never again shall we take on its load." In truth, our bodies are not burdens, we are not heavy laden because of them. We should not consider them servants or slaves.

It is the Spirit that knows the things of the Spirit; there is no man that can know the things of God save the spirit of God within him. The brain is not the cause of true thinking; the brain and true thinking are both effects of one Cause.

"When once it is perceived that the all is one spirit substance, the universe is seen to be within its source and cause; this is transfiguration and translation of form."

It is not by works that we are whole, lest any man should boast, for we, individually, are God's workmanship, created as sons and daughters, to all good things.

In the knowledge of the Truth that Divine Science teaches, we no longer see ourselves twain; there is not even a veil between the Holy Spirit and the body; we are one substance, we are free, forever free, in the. glory of at-one-ment with God.

Knowledge recognizes God to be the same at all times, that all there is, is God, the Creator and His Infinite Creation.

"Within the chamber of the Most High,
I ask the question, What am I?
And out of the stillness comes reply:
What is God? The same thou art;
He the whole, and thou the part."

224

Return to the Top   Return to the Top

PROVERBS. . - p. 225

THE BOOK OF PROVERBS.

Most representative specimen of the wisdom literature of the Bible. In harmony with the character of the Hebrew wisdom, which is inspired by religious motives, this book has a decidedly religious character, although we find also that many maxims have found their way into it which bear upon ordinary prudence of conduct, and are merely personal experiences.

Solomon is named as author of the Proverbs, but verse 6 appears also to announce the intention of publishing "words and riddles of the wise."	Prov. 1: 1
The Proverbs of Solomon form the real kernel of the book.	Prov. 10: 1
These "words of the wise" contain maxims and warnings which only exceptionally are comprised in a single verse; usually they extend to two, sometimes. three, once even to seven verses.	Prov. 22: 16
	Prov. 22: 17
This appears to be an appendix to chapters 22:17, 24:22, headed, "These are also words of the wise."	Prov. 22: 24
	Prov. 24: 23-24
With the heading, " These also are Proverbs of Solomon, which the men of Hezekiah, king of Judah, collected." 25:1. Here again, as a rule, each verse makes up a proverb. (Chaps. 28, 29.)	Prov. 25-29

Some of the sayings are duplicates of proverbs contained in 10:1, 22:16. These sayings are distinguished

225

by their concentrated force and the rich imagery of their language. The religious character recedes far into the background; particularly in Chapters 25, 27 they are for the most part, sayings bearing purely upon a prudent direction of the conduct of life.

Prov. 30:	Entitled "Words of Agur," made up of enigmatical sayings and a few numerical proverbs such as meet us in Prov. 6:6-19.
Prov. 31:1-9	Exhortations to Lemuel, king of Massa, spoken by his mother, who cautions her son against wine and women and exhorts him to rule righteously.
Prov. 31:10:31	

The book closes with a poem, standing by it self, without any connection with what precedes, devoted to a eulogy on the virtuous housewife.

It is a characteristic circumstance that these proverbs in the. main agree in their religious and ethical requirements with those of the prophets.

We find the same estimate of sacrifice in Prov. 15:8 and 21:3-27, as in Amos 5:18, Hos. 6:6, Ezek. 1:11 the same praise of Humility and warning against pride in Prov. 11:2, 14:29, 15:1, 4, 18, 25, 33, 16:5-8, 17:19, 18:12, 19:11., 21:4, 22:4, as compared with Isa.:11, Amos 6:8, Hos. 7:11 and Mic. 6:8; the same denunciation of those who oppress the poor, and the same commending of care for the latter in Prov. 14:31, 17:5, 18:23, 19:1-7, 22:2-7, 28:3-6, 27 29:13, as compared with Amos

226

4:1, Hos. 5:10 and Mic. 2:3-8. Like the prophets, these proverbs see in the fear (reverence) of God the foundation of all piety and morality, and in numerous passages they exhort men to this fear.

But the unique relation between Jehovah and Israel, of which the prophets never lost sight, has here disappeared., and the individual conception of religion has taken its place. The upright and the ungodly, the proud and the humble, the understanding and the foolish, are the contrasted categories with which the proverbs have to do.

In the proverbs monotheism (doctrine. of one God only) holds undisputed sway, and the consequences that result from it are not defended, but assumed as self-evident, and only the practical points of view insisted upon: God is the creator of rich and poor, Prov. 14:31, 22:2, 29:13; God is Omniscient, Prov. 15:3, 11, 16:2, 17:3, 21:2, 24:11 He directs all things, the actions of men, Prov. 16:1, 9, 33, 19:21, 20:24, 21:1, 30, 29:26.

There is no mention of the Messianic deliverance, but a belief in individual retribution, as this has been growing up since. the days of Jeremiah and Ezekiel. Prov. 10:3, 24, 29, 11:3, 8, 31, 12:2, 13, 21, 13:6, 15:29..

As in some of the Psalms, the godly are cautioned against the envy awakened by the prosperity of the wicked, and have their attention

227

directed to the righteousness of Jehovah. which will manifest itself in the future.

Return to the Top  Return to the Top

THE BOOK OF ECCLESIASTES

Represents a variety of earthly things, valuable in revealing the limitations of the Hebrew thinker who took neither immortality nor Idealism into account. Koheleth, "one who holds the office of teacher." By this name, the author means Solomon; he is identified with the son of David, king in Jerusalem. Ecc]. 1:1. It is also clear that Solomon is the king whose varied experiences of wisdom and luxury are referred to in chapters 1 and 2.

The title Ecclesiastes comes from the Septuagint, a Greek version of the Old Testament.

Eccl. 1: 2-11   There is no mention in the book, of Jehovah, God of the Jews. Vanity of Vanities, all is vanity. No profit comes to man for all his toil. Nature and man go ceaselessly round in the same course, and there is no new thing under the sun.

Ecc. 1: 12-15

Ecc. 2:13-14   Koheleth, being king over Jerusalem, uses his wisdom to understand the life of men and finds that all is vanity.

"I saw that wisdom excelleth folly, as far as light excelleth darkness --- yet I perceived that one event happeneth to the wise man and the fool." This "event" appears to be the time that man ceases his activity in this world, the wall of sense

228

to which Koheleth seems to be addressing his intellectual reasoning of the destiny of man and the value of his labor. One reasons from a basis, a principle one argues from appearances.

Although there is much pessimism, much that Is not true in this book, yet we find much to encourage us in what the author discovers in his search for the meaning of Life, which appears to give him so little comfort or satisfaction.

He (God) hath made everything beautiful in its time; also He hath set the world, or eternity, in their heart, yet so that man cannot find out the work that God hath done from   Ecc. 3:11-15 the beginning even to the end.

In the process of the soul's unfoldment into the consciousness of itself, the end or completion of God's perfect work cannot be seen, but the world or eternity set in man's heart is the light of understanding that analyzes the world belief and opinion, the intellect with its reasoning from appearances, and sees that all eternity belongs to the soul.

"I know that there is nothing better for them than to rejoice and do good as long as they live.   Ecc. 3:12

"And also that every man should eat and drink, and enjoy good in all his labor it is the gift   Ecc. 3:13 of God." God gave man life and work as compensation.

Ecc. 3:14

"I know that whatsoever God doeth, it shall

229

be forever; nothing can be put to it, that men should fear (reverence) before Him." American Standard Version.

A wonderful tribute to the God of Wisdom, Omnipresent, Omnipotent and Omniscient, who seems to be in the past, remains in the past to this debater upon the subject of man and his destiny. The fundamental thought of the book is, all is vanity, life yields no satisfaction.

Ecc.7: 29   "Behold, this only have I found: that God made man upright (equitable); but they have sought out many inventions."

Ecc. 9: 7-9   A great truth; the processes of Life are constantly changing as man slowly unfolds into the consciousness of his own individuality, his relation to God, a son of the Most High, endowed with all possibilities of bringing forth into expression the perfection that he is; a perfect part. of a perfect whole. The soul does not analyze, it realizes.

Koheleth urges a moderate enjoyment of the good things of life: "Eat thy bread with joy, and drink thy wine with a merry heart; let thy garments always be white and let not thy head lack ointment; live joyfully with the wife whom thou lovest all the days of the life of thy vanity." There is an inner freedom which can grow up alongside of all constraints of birth, custom, environment and ignorance.

When Koheleth speaks of God, we may easily

230

read more into his language than he meant, because he is seeing only surface conditions. His conception of God has nothing attractive or winning. He is rather set before us as the Omnipotent Ruler who has ordained all the course of history, which man vainly seeks to comprehend; an austere Deity on whose favor or forbearance none may venture to presume. When we learn to know that God is Law and that Law is Love (conscious unity), we may then realize that the use of God's law is not that of servitude, but it is freedom of fulfillment and self-revelation.

There was a little city and few men within it; and there came a great king against it and besieged it, and built great bulwarks against it. Now there was found in it a poor wise man, and he, by his wisdom, delivered it; yet no man remembered that same poor man.   Ecc. 9:14-15

Ecc. 17:18-19

The soul is this little city in which the strength of men, or human opinions, count not.

"Not by might, nor by power, but by my spirit saith the Lord of hosts." Zech. 4:6.

Charity and industry to be exercised in hope, he exhorts men to labor in joy, even amid uncertainties. "Money answereth all things." But seek ye first the kingdom of God and his righteousness and all these things shall be added. First things first.

From the viewpoint of the author of Ecclesiastes,

8a

231

there is no fundamental inconsistency; both life and death are indefinite. Now one, now the other, might, according to his mood, be esteemed the worse.

Vanity of vanities, saith the Preacher; all is vanity. We are gratified to note that this

Ecc. 12: 8	declaration is not the final one. We still have the Epilogue.
Ecc. 12:9-10	"And moreover, because the Preacher was wise, he still taught the people knowledge --he sought to find out acceptable words and that which was written uprightly, words
Ecc. 12: 12	of truth."
Jno. 14: 12	"My son, be admonished: of making many books there is no end; and much study is a
Ecc. 12: 13	weariness of the flesh." Our great Teacher left no written line; He lived the truth He taught, he also assured us of our possibilities to live the Truth by our adherence to
Ecc. 12: 14	Principle.

"Let us hear the conclusion of the whole matter: Fear (reverence) God and keep His commandments; for this is the whole of man." The wholeness of man is the consciousness of the presence of God; the expression in his life activity of the result of obedience to the commandments of the one who knew God as no man before Him; Jesus, a brother from the bosom of our own human family. Mark 12:28-35. Wisdom is justified of all her children!

"For God shall bring every word into judgment,

232

---

with every secret thing, whether it be good, or whether it be evil."

"Thou shalt call and I will answer thee; thou wilt have a desire to the work of thine hands. For now thou numberest my steps; dost thou not watch over my sin?" Job 14:15-16.

Return to the Top   Return to the Top

SONG OF SOLOMON.. - p. 233

THE BOOK OF THE SONG OF SOLOMON.

An allegory relating to the Bride and Bridegroom of the Spirit.

No one in Israel has ever doubted that the Song of Songs is a holy canonical book," says a devout Jew, about. the end of the first century A. D., "for the whole world is not worth the day on which the Song was given to Israel. For all the writings are holy, but the Song of Songs is a holy of holies."

Henceforward this idea of the incomparable value of the book continued to be the only prevailing one amongst the Jews, and thus it passed over also into the Christian Church.

One result of the allegorical interpretation was the introduction of the liturgical use of the Song into the Jewish church.

The Song of Songs --- Canticles, along with Ruth, Lamentations, Ecclesiastes, and Esther, made up the five "rolls" which were read to the congregation in the earliest times.

There was a Jewish regulation that no one

233

---

was to read the book till he was thirty years of age, the age, according to Num. 4:3, at which the Levite is ready to enter upon his sacred duties.

The Song of Songs, which is Solomon's:

Chapter 1. The Bride speaks with the daughters of Jerusalem.

Chapter 2. Loving converse between the Bride and Bridegroom.

Chapter 3. The Bride of the Spirit searches for its own expression.

Chapter 4. Dialogue in Oriental language of a lover speaking to his beloved.

Chapter 5. Bride of the Spirit speaking to the daughters of Jerusalem in the same language. Praise of the Bride and her response at the marriage feast. The Bridegroom's response.

Chapter 6. Temporary separation.

Chapter 7. Mutual praise of Bridegroom and Bride. Their union in the consciousness that God is Spirit, and that that which is born of Spirit is

Spirit; their eternal state of unity.

Sol. 8: 6, 7, 12, 13
Chapter 8. "Set me as a seal upon thine heart, as a seal upon thine arm" (strength), says the Spirit, for Love is as strong as death. Many waters cannot quench love, neither can the floods drown it. My vineyard, which is mine, is before me. Thou, O Solomon, shalt have the thousand;

234

---

thou that dwellest in the gardens (in the consciousness of the Spirit), the companions hearken to thy voice; cause me to hear it."

Make haste, my Beloved!

Sol. 8: 14

235

Return to the Top Return to the Top

Help us process more books to support your journey

Click here to give your support to New Thought Library

CHAPTER 11..

PROPHETS - p. 236

Prophet, "statesman in the kingdom of God."

Isa. 6

Jer. 1.

The prophets were not persons who stood as mere objective Divine instruments to the people whom they addressed; they were of the people; the life of the people flowing through the general mass reached its flood-tide only in them. And it was with hearts so filled and minds so quickened and broad that they entered into communion with God. Fountain of prophecy was communion with God. Holiness, nearness to God.

Prophets were taken from any class:

> Aristocracy of capital, Isaiah.
>
> Population of the country townships, Micah.

Prophets

> Those that followed after flocks, Amos.
>
> Priestly family, Jeremiah.

Women, too, might be prophetesses, Miriam, Deborah and Huldah.

2 Kings 22

The call came to the three great prophets through a vision --- Isaiah 6, Jeremiah 1, Ezekiel 1. Terms used in those days give no clear meaning for today.

Books of "Words," Amos 1:1, Hosea 4:1, Isaiah 2:1, Micah 1:1. Words or Message, or idea or purpose. These words were spoken rather

236

---

than written. Truth delivered through personality.

The prophets were first speakers, not writers.

The Books as they now stand were collected and arranged by disciples of the prophets, rather than by the prophets themselves. Just as Jesus left no written message, only by his disciples do we get his words.

Prophecy surpasses or transcends prediction, because it embraces a recognition of the cause as well as the effect.

Interpretation of the prophecies we are studying is largely dependent upon our understanding of principle, the Omnipresence of God, and of the laws of the figurative language of type and symbol. The subjects of prophecy differ but all are directed to one general end, the victory of righteousness and Truth over ignorance.

True prophecy is a Divine Message referring not alone to the future, but also to the present and past. It is the understanding of the law Of cause and effect. God, the Source and Cause all that has come forth, the effect.

Prophecy may be a revelation, a rebuke, an exhortation, a warning, a promise, or a prediction.

"I have also spoken by the prophets and I have multiplied visions and used similitudes, by the ministry of the prophets."

Hosea 12:10

The Old Testament prophets portray the coming

237

kingdom of righteousness and power, and then identify this highest good for humanity with a specific quality of being or fore-runner called the "Anointed One," which means the Christ principle through which this realization is attained.

We who spiritually interpret the writings of the Hebrews see man awakening from his dream of sense into conscious dominion and power of Spirit, which is the highest plane of consciousness. Therein he knows himself to be the temple of God and communes and worships in Spirit and in Truth. An unfoldment from ignorance into the light of understanding, from childhood into manhood.

The Scriptures must be given spiritual interpretation because the Spirit is back of the letter. "For the letter killeth, but the Spirit giveth life."

The prophets of Israel stood against the evils of their own people, loving righteousness better than the fatherland. They devoted themselves to bringing forth righteousness (right thinking) and in this they shine in the firmament of history more than the seers of other peoples. It was because of this that Christianity could find a foundation.

Return to the Top  Return to the Top

ISAIAH.. - p. 238

The name Isaiah signifies "The Salvation of Jehovah," and commentators consider it indicative of his high character. He was called the

238

"Prince of all prophets." He was the greatest of the Hebrew prophets, by the strength of his individuality, the wisdom of his statesmanship, the length and unbroken assurance of his ministry, the divine service which he rendered to Judah at the greatest crisis of her history, the purity and grandeur of his style, and the influence he exerted on subsequent prophecy.

Isaiah was born about 760 B. C., seven years before the reputed foundation of Rome he was a child when Amos appeared at Bethel, 755 or 750, and a youth when Hosea began to prophesy in North Israel. Micah was his younger contemporary. The possibility is strong that this prophet came of noble birth, from a royal family, and that he was a city-bred man. In interest and thought he .concentrates on the city his habits were adjusted to city life. He was influential in court; his usefulness and power as prophet were due largely to his knowledge of affairs.

Efficiency has place in the work of the kingdom of God. The length of Isaiah's career was at least fifty years. He lived through the reigns of Uzziah, Jotham, Ahaz and Hezekiah. Inscriptions of Assyria and Babylon relate deeds of monarchs for their glory. In Egypt laymen were buried with their own written statements of their good deeds for their gods to see. Very different these from the prophet's autobiography, which

239

---

consists of his visions of God and are not written in personal vanity.

Jerusalem is Isaiah's immediate and ultimate regard, the center and return of all his thoughts, the hinge of the history of his time, the summit of those brilliant hopes with which he fills the future.

He has traced for us the main features of her position and some of the lines of her construction, many of the great figures of her streets, the fashions of her women, the arrival of embassies, the effect of rumors. He has painted her aspect in triumph, in siege, in famine and in earthquake war filling her valleys with chariots, and again nature rolling tides of her fruitfulness up to her very gates; her moods of worship, panic and profligacy. It is for her defence that he battles through his years of statesmanship, and all his prophecy may be said to travail in anguish for her new birth.

What kindles the reason and style of the writer is the thought of God. Isaiah not only rebuked, but he inspired. The breadth and force of imagination, the daring treatment of the history of the world as a whole, may be traced to the writer's recognition of God's omnipotence, and are the signs of how absolutely he was possessed by this as the principle, the governing truth.

The most perfect apostle of Israel's monotheism,

240

---

no prophet is more daring in his ascription of passion to Jehovah; he does not hesitate to picture Him as an excited and furious warrior and as a travailing woman. Isa. 42:13-14. Israel's maker is Israel's husband. Isa. 54:5. It must be kept in mind that Jerusalem was the one spot on earth where Jehovah was worshiped. His shrine was there; there lived the only community which preserved for mankind the true knowledge of Him and His purposes, the little band of disciples to whom Isaiah committed his testimony and revelation.

Israel has turned backward. The prophet's was the voice that held the people to the religious spirit. Sodom and Gomorrah stand for destruction.	Isa. 1: 4
Sacrifice; rather obedience than sacrifices.	Isa. 1: 10
Cease to believe evil.	Isa. 1: 11-15
We shall see the purity in place of sin, ignorance.	Isa. 1: 16-17
Belief in more than one is their sin or mistake; duality.	Isa. 1: 18
Zion was a mount in Jerusalem, the highest ascent, highest vision. Jerusalem, consciousness veiled by conception. New Jerusalem, consciousness unveiled.	Isa. 1: 22

Not the personal but the dross beliefs. Zion shall be redeemed clear vision within us judges.

Isa. 1: 27

Isa. 1:28

241

Judges ruled from within kings ruled from without.

Isa. 2: 2-3	In the last days, the full understanding.
Isa. 3: 11	Personality debased, an idol, belief of anything outside of God.
Isa. 4: 1	Seven complete male and female, call of both natures to unite. 68th Psalm, 11th verse. "The Lord giveth the word the women that publish the tidings are a great host." A. R. V.
Isa. 5: 4	
Isa. 5: 13	Parable. Thoughts have run riot. Why, when we started right? We did not stay with our basis.
Isa. 6: 9-11, 13	Famine, lack of knowledge.
Isa. 7: 14	Withdrawing from God. Immanuel, God with us, Divine Nature. Spirit promised to man, not a personality. The sign, not referring to Jesus Christ principle, to any young woman who could see God with us, fulfilled in Ahaz' time, in less than a year.
Isa. 8: 7-8	
Isa. 8:19-20	Prediction of captivity of Israel to King of Assyria. King of Assyria represents belief of power against them.
Isa. 9: 2	
Isa. 9: 6	Should not a people seek unto their God? To know God, to understand the Law if they speak not according to this word, it is because there is no light (understanding) in them.
Isa. 9:14-17	
	Shadow, gloom we throw over death.

Prediction of the Christ Spirit in us, born to us.

Lesson to us, any leader that is personal. Impersonal consciousness is the Christ consciousness.

242

God's love a consuming fire, remnant shall lean upon Jehovah, the Holy One of Israel, in truth.	Isa. 10:17-21
When we know God there shall be no more hurt. Mark 16:15-19. Read 9th verse: waters, unlimited possibilities cover, absorb the sea. To the Israelites, Gentiles meant all other nations not God's people.	Isa. 11: 1-10
	Isa. 11: 15
Tongue, speaking --- sea, separation or limitation. Egypt, sense of duality.	Isa. 12: 3
"With joy shall ye draw water out of the wells of salvation."	Isa. 13:19-22
Destruction of Assyria, of that sense of enmity.	Isa. 14: 1-5
Divine Nature is there and cannot be lost. Staff --- what we lean on.	Isa. 14: 9-13
Lucifer fell, personality, belief in ignorance fell.	Isa. 14:16-17
No power in evil. "Resist not evil." Matt. 5:39. No power in evil to resist.	Isa. 17: 7-8

In that day, light of understanding, shall man look to his Maker, and not to the work of his own hands.

Isa. 17:10-11

Isa. 19: 1-3

Desolation that must come to our false hopes. Burden, dropping of our conceptions or purification. No spirit in the external except God. Ignorance shall destroy itself.

Isa. 19:18-19

Five senses are the avenues of Spirit to be

243

---

illumined, spiritual perception. Destruction of false vision or ignorance. Altar, where we acknowledge God in midst of externals.

Isa. 19: 25	Israel means Spirit; Assyria, active soul;
Isa. 21: 1	Egypt, external form; Spirit, Living Soul, Body.
Isa. 22: 1	Drop burden of the sea, sense of limitation, lack.
Isa. 22: 9-12	Burden of valley of vision, valley --- low vision.
Isa. 23: 1-2	Water of the old pool --looked not unto the Maker thereof, neither had respect unto Him that fashioned it long ago man looks to the external for source and cause.
Isa. 24: 1-3	Burden of Tyre, isle, separated from land or main consciousness, indicates sense of separation.
Isa. 24: 6-9, 19, 23	Tyre stands for those not of Israel, or spiritual nature.
Isa. 26: 4-5	Earth in material sense. Third verse, land, two meanings, consciousness and our sense of belief. Destruction of the earth or negative belief. Brings down personality.
Isa. 27: 13	When we cease beliefs.
Isa. 28:16-18	The Plowman.
Isa. 28:23-26	Rebuke for those who do not live sincerely.
Isa. 29:13-14	"Sit still" means to be established.
Isa. 30:1-3, 7	Light of perfection.
Isa. 30:25-26	Third verse explains verse one. Leaning on the external, shall fall.
Isa. 31: 1-3	Man's infinite possibilities.
Isa. 32: 2	Cannot dwell in Zion and know any iniquity.
Isa. 33:20-24	

244

Destruction of conception of heaven and earth. Eighth verse, God's vengeance and redemption.

Isa. 34:1-2, 4, 8

We do not see desert, but Infinite possibilities.

Isa. 35: 1

Same as II Kings, 19th chapter.

Isa. 37:

Same as II Kings, 20th chapter.	Isa. 38:
Isaiah, not God, had said, "Let them take a lump of figs and lay it for a plaster upon the boil, and he shall recover."	Isa. 38: 21
John the Baptist's call to prepare the way for Christ consciousness in each one of us.	Isa. 40: 1-5
The Lord is calling on claim of human senses, asking them to tell what they can do.	Isa. 41:28-29
"I Am with thee, fear not."	Isa. 43: 2
"You will see only My Presence in all the empty places. I Am the first, I Am the last and beside Me is no God."	Isa. 43:18-19 Isa. 44: 6
Darkness is full of God, God gives us omnipresence, but until we see it, it is secret. God is all, the I Am.	Isa 45: 3
None but One, God.	Isa. 46: 9-11
"When I call unto them, they stand up together" --- a unit.	Isa. 48:13
Voice calls within, when we have sense of separation, isles, listen; the thoughts that have wandered from God are called to know the Truth.	Isa. 4 9: 1 Isa. 49:14-16
"In Him we live and move and have our being."	Isa. 50: 1-4
We are punishing ourselves.	Isa. 50: 11

A choice, but Law follows,

245

Isa. 51: 3, 6, 9, 21	Awake, put on strength, thou afflicted and drunken, but not with wine (with ignorance). Ignorance has no cause.
Isa. 32:1-3-4	Holy One of Israel, Divine Consciousness.
Isa. 54: 5	Blindness on our part.
Isa. 54: 7 8-13	Beautiful assurances.
Isa. 57:20-21	There is no peace to the wicked, the ignorant.
Isa. 58: 1-2	Voice that came to the prophet, showed the people their wrong thoughts.
Isa. 58: 5-9	True fact and promise. 12th verse, not a reward, but the Law. Repairer, when you see or know there is no separation.
Isa. 59: 1, 2, 20, 21	Eternal covenant.
Isa. 60: 1-2, 18-22	Description of Zion's glory, the highest realization of the soul.
Isa. 61: 1-4	Supposed to refer to the office of Jesus. Office of the one that sees the Christ and accepts it.
Isa. 62: 1-4	A new name. Rev. 2:17. Fourth verse, change indicated by these names is that of a female slave, who is set free by her master and then married by him. Emancipation of soul from bondage and its establishment in its true relationship with its Source.
Isa. 63: 4	
Isa. 64: 4	

Classing together of vengeance and recompense, compensation.

Isa. 65: 1   What God hath already prepared for his children. II Con 11:9.

Until we seek God.

246

Blessed state of New Jerusalem. Dust shall be the serpent's food.

So, many conceptions to be slain.

The Eternal.

Isa. 65: 17 20-25

Isa 66 7-9, 12, 15, 16

Isa. 66: 22

Return to the Top   Return to the Top

JEREMIAH.. - p. 247

Jeremiah, "Exaltation of the Lord," was born of a priestly family in Anathoth, a small village close to Jerusalem, and prophesied from the 13th year of Josiah till after the captivity, a period of forty years, all through the Babylonian invasion.

The conflict in his mind reveals the duality in his thought characteristic of his whole life. Personality and individuality wrestle within him no less than they do in St. Paul. He was subjected to cruel persecution by the rules of Jerusalem --- his warnings were neglected, but fulfilled; his fellow citizens carried away captive and Jerusalem a heap of ruins; in an adjoining cave he wrote his Lamentations over it.

Chapter 43, a remnant rallied around him after the murder of Gedaliah and were forbidden by God through his words to flee into Egypt; but they accused him of falsehood and carried him with them into that country, where, according to Jerome, he was put to death. Formula of his prophecies, " The word of the Lord came to Jeremiah."

Chapter 1, an introduction, probably prefixed to the whole at the final revision.

247

Chapters 22 to 25, shorter prophecies delivered against the kings of Judah and false prophets.

Chapters 25 to 28, two great prophecies of fall of Jerusalem.

Chapters 29 to 31, message of comfort for the exiles of Babylon.

Chapters 32 to 44, history of the last two years before the capture of Jerusalem and of Jeremiah's work during that and the subsequent period.

Chapters 46 to 51, prophecies against foreign nations ending with the great predictions against Babylon.

Chapter 52, the supplementary narrative, which is also a preface to Lamentations.

Jeremiah, when called to the prophetical office, pleaded his youth.

Jer. 1: 7

Jer. 9: 1

Is anyone called or chosen? No, the choosing is on our part, the chosen are those who choose.

The prophet was considered inferior in elegance to Isaiah. He predicted calamities and grieved over the judgments to befall Israel. Yet, to him, backsliding Israel was justified above treacherous Judah. In Jer. 13:25 he gives the reason for his people's condition, "Because thou hast forgotten me and trusted in falsehood." Is not that reason enough? Forgotten the Source, our very life, and trusted in that which is not true.

248

The new covenant means the wiping out of the old covenant. God has no new covenant, but through being perfected in love we are more willing to hear the voice of God and it has a different meaning to us. When we hear it in a new way, a new light, we call it a new covenant. It is new to our consciousness. We hear it and obey it the best we know. People say, "If we do the best we can, why do we suffer?" A child at school works an example wrong, gets the result of it. Would you say, "She has tried hard why not let her pass?" You cannot!

Jer. 31: 31

We are hearing a new covenant right now and are obeying it it bears a higher message to us, not that God has any higher message, but we hear it in a higher or clearer consciousness.

Divine nature within each. We are all priests of God.

Jer. 1: 3-5

Consecrate my words. Indicative of the right attitude. Arise, stand on your feet.

Jer. 1: 6, 77, 9, 10,

Rebukes Israel is not innocent.

"Yet I had planted thee a noble vine, wholly a right seed; how then art thou turned into the degenerate plant of a strange vine unto me?" The house of Israel is ashamed! Saying to a stock, "thou art my father"; to a stone, "thou hast brought me forth." But in a time of trouble their kings, princes, priests and prophets will say,

Jer. 2: 3, 5, 6, 7, 13, 17, 18, 19

249

Jer. 2: 21, 26-28, 35-36,

Jer. 4: 1, 2, 4

Jer. 7:22, 23

"Arise and save us." But where are thy gods -- let them arise, if they can save thee in the time of thy trouble. Thou also shalt be ashamed of Egypt, as thou wast ashamed of Assyria. We always turn to God in our extremity.

Jer. 8: 8-22

Jer. 11: 5-8

Nations, everybody outside of Israel. Circumcise, cut off belief in the external as cause.

Jer. 12:12

Jer. 15: 16

God's command not concerning burnt offerings or sacrifices, but "Obey my voice, and I will be your God, and ye shall be my people. Walk ye in all the ways I have commanded you, that it may be well unto you."

Jer. 17:12-14

Jeremiah's lament.

Jeremiah proclaims God's covenant, "Hear ye the words of this covenant, and do them."

Flesh used as sense. Telling the Law not a curse. Making flesh his strength. It is the Spirit that quickeneth. One who trusts in God, sees.

"Thy words were found and I did eat them; and thy word was unto me the joy and rejoicing of mine heart: for I am called by thy name, O Lord God of Hosts." This is the cause of his rejoicing. My name is my nature; called by the character of God. When we drink in these words and realize them, that is "eating the words."

A glorious high throne from the beginning is the place of development, because they have forsaken the Lord, the fountain of living waters,

250

Source of infinite possibilities. "Heal me, O Lord, and I shall be healed."

Meaning of God's repentance, which is really we who change. Leave my fields for a rock of the field. Shall the running waters be forsaken for the strange cold waters? John 4:14.　　Jer. 18: 8-14

"I will gather the remnant of my flock; I will set up shepherds over them which shall feed them and they shall fear no more, nor be dismayed, neither shall they be lacking," saith the Lord. I will raise unto David a righteous (right thinking) Branch, who shall execute judgment and justice in the earth. His name shall be called Jehovah our righteousness. A. R. V.　　Jer 5, 4, 36, 23: 1, 4,

23rd verse: "I, a God at hand, and not a God afar off." 24th verse: "Do not I fill heaven and earth?" 36th verse: "Every man's own word shall be his burden, for ye have perverted the words of the living God."

God's repentance, when ye turn.

This man is not worthy to die, for he hath spoken to us in the name of the Lord our God.　　Jer. 26: 2-3

"And ye shall seek me and find me, when ye shall search for me with all your heart."　　Jer. 26: 8-9, 11-16

Deliverance from captivity promised, whirlwind of one's own thoughts.　　Jer. 29: 1, 10-15

Every man that eateth the sour grapes, his teeth shall be set on edge. They shall say no more, "The fathers have eaten a sour grape and　　Jer. 30: 5-8, 10, 22, 23

251　　Jer. 31:29-32

---

the children's teeth are set on edge," but every one shall die for his own iniquity, shall die to his own ignorance. "I die daily" --- Paul.

	Jeremiah Imprisoned.
Jer. 32: 2	
	Promises of God to all his children.
Jer: 33: 6-7	
	Anger of the Lord --- God a consuming fire.
Jer. 36: 7	
	The voice of condemnation, shutting up this voice, holding it within us, always shuts
Jer. 37:17-21	us in prison.
Jer. 38 .	Had to sink in mire, an attitude full of condemnation, miserable sinner, worm of the dust.
Jer. 39: 11, 12, 14	
	King Nebuchadnezzar protects him. "Do unto him even as he shall say unto thee," he
Jer. 42:15-17	tells the captain of the guard.
Jer. 43: 4-7	God's word to the remnant of Judah.
Jer. 44:26-28	Return to Egypt.
Jer. 46: 11-17, 19,	And all the remnant of Judah, that are gone into the land of Egypt to sojourn there,

20, 24-28   shall know whose words shall stand, mine or theirs. Yet a small number shall return.

Jer. 48:46-47   "0 Virgin, daughter of Egypt, in vain shalt thou use many medicines; thou shalt not be cured. Pharaoh, king of Egypt, is but a noise. Furnish thyself to go into captivity --- destruction cometh. The daughter of Egypt shall be confounded. I will not make a full end of thee, but correct thee in full measure; yet will I not leave thee wholly unpunished."

Judgment and restoration. God's vengeance

252

---

the starting point for redemption, God's judgment the beginning of restoration.

All that Babylon stands for (that which takes us into captivity) shall be wiped out.

Jer. 50: 2-6

Claim of separation has come over Babylon.

Jer. 51: 8, 42, 64

Thus far are the words of Jeremiah.

Jer. 52:

Same as Chapter 22 of II Kings.

Return to the Top   Return to the Top

Help us process more books to support your journey

Click here to give your support to New Thought Library

LAMENTATIONS. . - p. 253

A pathetic ode, full of tenderness, expresses Jeremiah's grief for the destruction of Jerusalem and its temple. Five distinct poems. Its original Hebrew title was "Ekah" (How), which is its first word, and was a common prefix to a song of wailing.

When we read it in the true light, we can get joy out of it. We see what God is destroying; and if Jeremiah lamented it was because he thought it was the anger of the Lord that was the destructive power.

The book is nearly all summed up in the fifth chapter, the epitome.

Return to the Top   Return to the Top

Don't get devoured by the costs

Get an accurate picture of New Thought Education

DivinitySchool.net

Ezekiel, "Strength of God," was a son of Buzi, a priest carried captive with other nobles by Nebuchadnezzar (B. C. 599) before the destruction.

He was settled with a Jewish colony on the banks of Chebar, two hundred miles north of Babylon.

253

He did not begin to prophesy till the fifth year of his exile.

Prophecies, two parts:

I. Those spoken before the destruction of Jerusalem to disabuse the' people of all false hope of help from Egypt., instilling into them the certainty of God's vengeance, and exhorting them to repentance.

II. Full of consolation, exciting hopes of future restoration on their true repentance, and the final glory of God's blessings upon them and the future resurrection of the flesh as flesh.

Between these two parts is a middle portion, denouncing God's judgment on the seven heathen nations around them. This was written between the commencement of Nebuchadnezzar's siege of Jerusalem and the news of its fall.

Hebrew tradition asserts that Jeremiah and Ezekiel exchanged writings in their lifetime, so that those of the former were read in Babylon, and those of the latter in Jerusalem.

The great obscurity of the book, from its allegorical form, and the apparent discrepancy between it and the Hexateuch --- Ezek. 18:20 and Exodus 20:5 led the Jews to place it among the "Treasures," which no one might read before the age of thirty, and for the same reason the Sanhedrin hesitated to give it a place among the canonical books of the prophets for public reading

254

---

in the synagogue. There are no direct quotations from it in the New Testament, although in Revelation there are several allusions and parallel pas- sages which show that it was known to the writer. Ezekiel introduces himself in the first chapter, 1:1-3, he was contemporary with Jeremiah and Daniel.

As the result of his labors the people improved in character toward the close of the captivity. They ceased from the worship of idols and re- turned to the worship of God. Because of his beautiful conceptions and extensive knowledge, he has been compared with Homer.

Gives the duties of a prophet. We have an idea that a prophet only foretells events, but his office includes much more than that. A prophet is a watchman, to watch over and   Ezek. 3: 16-21 warn Israel of sin and to offer rewards. This is all on a low plane of unfoldment, but while we sin, or think we can sin or not sin, this contains a lesson for us.

Ezekiel's prophecies are given in visions and strange images. " The heavens opened and I saw visions of God." The first vision is a whirlwind from the north, a great cloud and a fire. (There is always fire and light in these visions.)

First vision a peculiar figure. Only in the transparent atmosphere of the presence and power of God can Ezekiel's vision be understood.

Out of the north, "a fire unfolding itself, " and

9

out of the midst "the likeness of four living creatures; they had the likeness of a man, and everyone had four faces." The central face, man individuality; on the right (constructive) side the lion, strength; on the left (negative) side, the ox, sacrifice; the eagle, insight, wings joined, unity in purpose as they move with the spirit straight forward.

"As for the likeness of the living creatures, their appearance was like burning coals of fire, like the appearance of lamps; it went up and down among the living creatures; and the fire was bright and out of the fire came forth lightning." God's love a consuming fire; "flashes of lightning," the illumination of universal intelligence, "In the hour when ye think not, the Son of man cometh." Matt. 24:44.

Ezek. 1:15-24 Ezek. 1:24-25	Wheels within the wheel, circles within, a circle, man within God, the universal including the individual, the microcosm in the macrocosm. Man's work is in this circle, unlimited in boundary; man's work unlimited in his divine possibilities, "full of eyes."
Ezek. 1:26-27	"The noise of their wings, like the noise of great waters, as the voice of the Almighty" made him feel the power of God. "And there was a voice," he tells of it later.

The appearance of a man above (the throne) upon it. And I saw in the translucent glow the

appearance of his loins --- even upward --- even downward, the appearance of fire and it had brightness round about. The prophet's vision of glorified man. Zech. 1:28. The glory of the Lord. " This was the appearance of the likeness of the glory of the Lord; I fell upon my face and I heard a voice saying, "Ezekiel, son of man, stand upon thy feet (understanding) and I will speak to thee." It has been said, "As a prince you have power with God." We are to stand on our feet, not to grovel. Ezra was also addressed as the son of man, so Jesus was not the first to be addressed in that way.

Second vision. "Eat this roll and go speak to the house of Israel." Jer. 15:16 and Rev. 10:9. Unfold it into the consciousness of the word that was with God and was God.	Ezek. 3: 3-4
Right attitude, feet, understanding. We stand in God.	Ezek. 2: 1
The work of a prophet; prophet has a little clearer vision; the thought is to follow; watches over thought.	Ezek. 3: 17
	Ezek. 6: 8
Remnant, the Divine in us.	
	Ezek. 13: 2-4
False prophets reproved.	
	Ezek. 14: 20
Ranking Daniel with Noah and Job.	
	Ezek. 16
By righteousness (right thinking) each man saves himself. Jerusalem's unfaithfulness and her punishment, yet God remembers His covenant.	Ezek. 18: 2-3

Even under the old dispensation, the curse of

heredity is removed,. the inheritance of evil, or ignorance.

Ezek. 18:20-25	"The soul that sinneth, it shall die." "I die daily," says Paul. Die to sin or ignorance. "Is not my way equal?"
Ezek. 18: 32	"Turn yourselves," "Be ye transformed by the renewing of your minds --- and live."

Ezek. 20: 5	"I Am the Lord your God."
Ezek. 21:26-27	"Exalt him that is low, abase him that is high --- I will overturn --- this also shall be no more until he come whose right it is and I will give it him." (The Christ Spirit.)
Ezek. 22: 18	
Ezek. 20: 2-3, 6, 9	The fire, God's fury, consuming fire which purifies. Ezekiel's wife dies. The strength of his faith in God shown Judah by the prophet at this time. All who were with him had relatives in Jerusalem and death would come to many of them; they must not give way to personal grief in "the day when I take from them --- the joy of their glory that whereupon they set their minds, their sons and their daughters." "In that day (understanding) thou shalt be a sign (of strength) unto them,
Ezek. 29: 10, 13, 14	

and according to all he hath done shall ye do and they shall know that I am the Lord."

Egypt, our thoughts of the external, rivers, active possibilities. 9th verse, personal opinion claiming to have made its own possibilities.

Forty years, process of fulfillment.

258

Fire, God's love consuming all Egypt. Trust in Spirit, not flesh. The flesh profiteth nothing of itself, it is the Spirit that quickeneth 22nd verse, belief in personal power will be broken up.	Ezek. 30: 8, 21, 22
Assyrian, outward. Israel, turned inward.	Ezek. 31: 3-4, 7
Truth annihilates ignorance.	Eze -9k. 31: 8
The external looks so tempting. 10th verse, personal pride. 11th verse, therefore, it was rooted in the personal.	Ezek. 34:
	Ezek. 36: 22-24
Unfaithful shepherds of Israel condemned. 10th verse, therefore, I am against the shepherds, leaders. 31st verse, sheep stand for Spiritual Nature. Jesus teaching, this. Redemption of Israel. The Law says, I am not doing this for your sake, but because my Divine Nature is in you.	Ezek. 37: 1-14
	Ezek. 39: 23-24
Valley of dry bones. Dry land needed water, bones needed life. If bones are raised up in consciousness, the rest will be. These are lessons of God's power. 24th verse, Spirit of David, impersonal. God seeks those that worship in an impersonal way. David, the Beloved, Our Shepherd, one. Prediction of ingathering of all in unity. 28th verse, conscious realization of Omnipresence.	Ezek. 4.7:3-5
	Ezek. 48: 35

Israel to be restored.

Vision of holy waters. Ezekiel is shown possibilities of God.

Touch of Omnipresence. City, number of houses, house, individual consciousness; name of

259

---

city, Omnipresence. Ezekiel believed in the kingdom of the Spirit. His was the voice from the deep place in man's soul when inspired by the strength and the faith of the Spirit. The light "shining out of the darkness," "the light that lighteth every man that cometh into the world."

Return to the Top   Return to the Top

DANIEL. - p. 260

Daniel means, "God is my judge."

First six chapters historical, the last six chapters prophetical.

Dan. 1: 3-7 Dan. 1:17-21	Daniel was a youth of noble descent and of high physical and mental endowments. He was carried by Nebuchadnezzar in the third year of Jehoiakim from Jerusalem to Babylon and with other Jewish youths, especially three companions, Hananiah, Mishael and Azariah, assigned for education at the king's court.
Dan. 2: 16	They excel in body and spirit.
Dan. 2: 22, 30-34	"Daniel went in and desired of the king that he would give him time, and that he would show the king the interpretation."
Dan. 3: 1	Very impersonal. The dream, 31st to 36th verse. Duality. Feet, understanding. Stone, means Christ, Truth. Claim of mixed or divided understanding will be broken.

The image Nebuchadnezzar set up. We all pass through the purifying fire.

260

Loose --- not bound.

Dan. 3: 25

Belshazzar's feast and the writing on the wall.

Dan. 5:

The Queen advises him to send for Daniel, who interprets the writing. 23rd verse, "Thou hast lifted up thyself against the Lord of heaven --- and the God in whose hand thy breath is, and whose are all thy ways, hast thou not glorified." Darius, having taken the kingdom after Belshazzar, was slain, appointed 120 princes over whom he placed three presidents, making Daniel the first. These jealous princes persuaded the king to pass a decree that no God should be invoked for thirty days. Daniel was disobedient and was cast into a den of lions.

We do not find Daniel compromising with ignorance.

Dan. 6: 16

His restoration.

Dan. 6: 22

Decree of Darius.

Dan. 6 26-28

The last six chapters give his visions and prophecies.

Dan. 7: 3

Four beasts rise out of the sea. Interpreted.

Compare Dan. 7:3 and Revelation 13:1.

Compare Dan. 7:9 and Revelation 1:14.

Compare Dan. 7:10 and Revelation 5:11 and 20:4-12.

Compare Dan. 7:11 and Revelation 19:20.

Compare Dan. 7:13 and Revelation 1:7-13 and 14:14.

261

Dan. 8	Vision of rain and he-goat. Gabriel interprets the dream. Daniel sees a glorious. The man comforts him.
Dan. 10: 7	Last chapter, prophecies of the fall and rise of Israel and other nations the restoration
Matt. 24: 15	of the Jews and the establishment of the final universal kingdom of Christ Jesus

speaks of Daniel as a prophet. An allusion is made to him. Heb. 11:33-34. His writing not strictly prophetic. It has a far wider range, disclosing philosophy of history revealing to the Jews the great mission destined for them in the regeneration of mankind.

Thus, with. Ezekiel, the latter portion of the book of Daniel forms the connecting link between the prophecies of the Jewish dispensation and the more universal revelation of Jesus and his followers.

Daniel was supposed to have died at Sheshan, Persia, about 500 B. C., age near 94 years. His advanced age being the probable reason why he did not return to Judah with other Hebrew captives under proclamation of Cyrus (Ezra 1:1) B. C. 536, which marked the end of the 70 years captivity.

262

Return to the Top Return to the Top

CHAPTER 12.

HOSEA. . - p. 263

Prophets

HOSEA.

Hosea, Joel and Amos were contemporaries. It is probable that Joel prophesied to Judah at the same time that Amos forewarned Israel, and that these two slightly preceded Hosea, who like Amos was sent to the Israelites.

Hosea means "Saviour or Safety." These names suggest an exalted spiritual consciousness. Idolatry of Jeroboam had produced all kinds of vice, kings were profligate, priests had introduced shameful rites throughout the land, God was forgotten.

The rulers looked to Assyria for help and Hosea compares their defection to the unfaithfulness of a wife to her marriage vows his illustrations are taken from the rural and domestic pursuits and give insight into the modes of the life of that day. He reproves the Jewish people generally, but the Israelites especially, for idolatry. The message of all the prophets was almost the same, but often expressed differently. They nearly all end with the promise of deliverance. To read these writings from the old orthodox viewpoint it sounds as though God was constantly

9a

263

changing; at one time saying, "I will not have anything more to do with you"; then "I cannot forget my promise." Of course it was not God, but the people who changed.

Hos. 1: 7	How often it is said that the horses and chariots shall fall. "Woe to them that trust in horses" means, woe to them that put their faith in the strength of the external instead of in God. In connection with Hosea 1:7, read Zech. 4:6 and 9:10, Isaiah 1:21, Micah 5:10. It means trust not in externals.
Eph. 6: 10	
1 Cor. 6: 14	That is the true strength.
1 Cor. 15: 24	"And God hath both raised up the Lord, and will raise us up by his power."
Hos. 2: 1	"When cometh the end, when he shall have delivered up everything to the Father."
Hos. 2: 5	Ammi, "my people" and Ruhamah, "having obtained mercy."
Isa. 1: 21	A description of how the children of God have gone after other helpers than the spirit of God.
Hos. 2: 6	
Job 19: 8	Here this harlot is called "the faithful city."
	The city that becomes a harlot is the mother that seems to have departed after false Gods.

"I will hedge up thy way with thorns." When we feel the prick of thorns, we think it is cruel, but it is the Spirit trying to keep us from going after strange Gods.

The same thought, when we depart into ignorance

264

the way is hedged so that we cannot go too far in it or be completely lost.	Hos. 2: 8
"For she did not know that I gave her corn and wine and oil." She did not know that I, God, did it!	Ezek. 16: 17-18
See the departure from God. He has given us all these things, and after having received the supply for every need, we have forgotten it and have turned to the external for our help.	Hos. 2: 9, 11, 13
	Hos. 2: 14
Who are our lovers when we forsake the Lord?	Hos. 2: 16
They are the things we love in the external, the effect, that make us forget our source.	Hos. 2: 17
This is what is going to be done; where is the wilderness? Where there is no corn, or flax or jewels. It brings her just where the prodigal son was brought, where there was nothing but starvation.	Hos. 2: 19-20
	Hos. 3: 4-5

Baal-i means "My Lord," and Ishi means"husband."

Hos. 4: 6

Baal-im means "false gods."

Hos. 4: 15

Is not that a beautiful summing up? Have we not reason to feel that there is no such thing as being rejected by God?

Hos. 4: 17

An ephod was part of the priest's dress; teraphims were images.

"My people are destroyed for lack of knowledge."

Sounds as though Judah were favored. Anything we put in God's place is an idol.

265

Hos. 5: 13

"Yet could lie not heal you, nor cure you of your wound."

Hos. 5: 15

"I will wait until they turn to seek me." Affliction comes through disobedience to the Law, and through that the soul turns to God. Ps. 72:12.

Hos. 9:10-17

"According as they loved." Israel was in captivity to Egypt because it became

Hos. 10: 1-2

captivated by Egypt, by externals. The lovers I go out to get my supply from are the things I love more than God. 17th verse refers to Israel and the ten tribes; they are

Hos. 10-13

supposed to be these wandering tribes.

Hos. 11: 1

Their heart is divided. "You have taken the things I have given you and have made images out of them." God has given the things, but man has made images of them.

Hos. 13: 2

Dent. '28:64. These wandering tribes are referred to.

"Ye have ploughed wickedness, ye have reaped iniquity, ye have eaten the fruit of lies, because thou didst trust in thy ways, in the multitude of thy mighty men."

"When Israel was a child then I loved him, and called my son out of Egypt." There is what we have been studying. Every one of us, as the son, is called to come out of Egypt, out of externals. Egypt was all right, but it was not made to be a source to us.

"And now they sin more and more, and have made them molten images of their silver, and idols

266

---

according to their own understanding." Having the understanding darkened, being alienated from the life of God, through the ignorance that is in them because of the blindness of their heart. Gal. 4:8-9, Rom. 1:25.

No God but Me, no Saviour beside Me. "The calves of our lips," the first fruits of our lips, the very best, the praise offering.

Hos. 13: 4

Ephraim means fruitful. The constant re proof here is to Ephraim, that should bring forth fruit and does not. We should be wedded to God, our Source, that all we show forth would be fruit unto God.

Hos. 14: 1-9

Hos. 14: 8

Return to the Top  Return to the Top

JOEL. . - p. 267

Joel means "He that wills, or commands."

Joel was of the tribe of Reuben, son of Pethuel, and lived at the time of Uzziah.

This book is quoted by Peter, Acts 11:16-21, and Paul, Rom. 10:13.

Joel gives the usual admonitions, and urges repentance. He more than the others declares the punishment of the enemies. We find a great deal in this book about the "day of the Lord."

This means the destruction of ignorance, light destroying darkness.

Personal opinions and beliefs shall be destroyed.

Joel 1: 15

Isa. 13: 6-9

267

Jer. 30: 7

He, himself, shall be saved out of it. The Self is always saved.

Joel 2: 1

You see the inhabitants of Zion do not tremble, but the inhabitants of the land tremble. The personality in us trembles before the Truth. Isa. 13:10. When the greater light of the Lord's day comes, we shall not need the sun, moon and stars. Matt. 24:29. We rejoice in this, "The stars shall fall" means that we shall drop all personality, we shall be individuals, but not as light apart from the Great Light.

These are all prophecies.

Acts 2: 20

We rejoice in the day of judgment and we know that it is at hand, now, literally.

Rev. 6: 12, 14, 17

Reverence for day of later understanding. Preparation for the great day.

Joel 2: 21, 23, 27, 28, 32 (Read)

"The day of the Lord is near in the valley of decision."

Joel 3: 10

"Beat the swords into plowshares." This is afterward, when we find there is nothing to resist.

Isa. 2: 4

Judah to be delivered.

Joel 3: 14

Joel 3: 19-21

Return to the Top   Return to the Top

Amos means "waiting."

Amos prophesied to the ten tribes within the twenty-five years during which Uzziah and Jeroboam were contemporary two years before the earthquake. Zech. 14:5. He was, until sent by

268

God to prophesy at Bethel against worship of calves, a shepherd of Tekoa in Judah. He must also have preached at Samaria, since he rebukes the vices of a capitol, extreme luxury, revelry and debauchery, and contrasts them with the excessive poverty and oppression of the poor.

Israel was at the height of its prosperity under Jeroboam.

Amos preached against the nations around the two kingdoms (Israel and Judah), Syria, Philistia, Tyre, Edom, Ammon and Moab. He describes the condition of Israel and Judah, and especially charges Ephraim with ingratitude and obduracy.

Being a herdsman, he was not educated at the school of the prophets. He shows Israel how their iniquities bring punishment upon them. In the fourth chapter the Lord tells what he has done for them --- "yet have ye not returned to me." He repeats this five times.

Amos 7: 14-15

Amos 2: 10-11

Amos 4: 12

"I brought you up from the land of Egypt, and led you forty years through the wilderness to possess the land of the Amorite. And I raised up of your sons for prophets, and of your young men for Nazarites. Is it not even thus, 0 ye children of Israel? saith the Lord."

Amos 5: 4

"Therefore, thus will I do unto thee, 0 Israel; and because I will do this unto thee, prepare to meet thy God."

"Seek me and ye shall live."

269

Amos 5: 15

Amos 8: 11

Amos 9: 11, 14-15

"Hate the evil and love the good." We do not hate anything, not even ignorance. Rom. 12:9.

This is not a much higher statement. "To depart from evil (or ignorance) is understanding." We never get rid of evil by hating it we are to give up the belief in it as a reality.

The real famine.

"In that day will I raise up the tabernacle of David that is fallen--I will bring again the captivity of my people of Israel; they shall build the waste cities and inhabit them, and I will plant them upon their land, and they shall no more be pulled up out of their land which I have given them, saith the Lord thy God."

Return to the Top   Return to the Top

OBADIAH. . - p. 270

Obadiah means "servant of the Lord." He prophesied before the destruction of Jerusalem and conquest of Edom. As Nahum foretold the downfall of Assyria and Habakkuk that of Chaldea, so Obadiah predicts that of Edom.

	Very similar to passage in Jeremiah 49:14.
Ob. Chapters 1 to 8	The Edomites fancied themselves secure in the fastnesses of their rocks.
Ob. 1: 3	The spoiler should utterly destroy them.
Ob. 1: 4-16	The deliverance of Judah. All those who are saviours. "And saviours shall come upon Mount Zion to judge the Mount of Esau; and the kingdom shall be the Lord's."
Ob. 1: 17-21	
	270

Return to the Top   Return to the Top

JONAH. . - p. 271

The oft-recurring symbolic way of writing the name of the city, Nineveh, characterizes it as Ninu-a or Ni-na-a, "fish-dwelling." Perhaps from this sprang the tradition of Jonah and the whale. The name Jonah signifies a "dove."

Jonah, son of Amittai, was born at Gathepher in Zebulun, two miles from Sephorim. He is the same prophet that is sent to Jeroboam (II Kings 15:25) in answer to the bitter cry of affliction that rose from Israel. The deliverance there worked by God brought no return of allegiance to Him.

Jonah is next sent with a message of warning to the Ninevites, but disobeys the command. The clue to his unwillingness and murmuring may have been his foreknowledge that the nation so spared was destined to be God's instrument for punishment of his native country. He embarks at Joppa to flee to Tarshish (4:2), the most remote quarters of the earth to which the exiled Jews may have fled.

His experiences are told in symbolic language of the Orient. The tempest in his own mental realm, between his personal desire and the Divine command, resulted in his wonderful prayer. The last lines, "When my soul fainted within me I remembered the Lord; and my prayer came in unto thee, into thine holy temple. They that observe lying vanities, forsake their own mercy. But I

271

---

will sacrifice unto thee with the voice of thanksgiving; I will pay that that I have vowed. Salvation is of the Lord."

When the command was repeated, Jonah arose and went unto Nineveh and cried unto the people "Yet forty days, and Nineveh shall be overthrown." Jonah kept his vow, unwillingly, but the people of Nineveh believed him and proclaimed a fast and put on sackcloth, from the greatest of them even to the least of them. And God saw their works, that they turned from their evil way; and did not do unto them that he had said.

Jonah cannot release himself from his personality, as he sits in his booth, "till he might see what would become of the city." He was pleased when the vine grew up over his booth to shade him, but when the worm destroyed it and the sun and wind beat upon his head, he fainted.

"Doest thou well to be angry for the gourd?" In Jonah's mental condition he was angry with the gourd and everything else. He felt that God was making a great mistake in saving these people. It made him lose all interest in life. Divine Love says to him, " Thou hast had pity on the gourd, for the which thou hadst not labored, neither madest it grow --- and should not I spare Nineveh, that great city, wherein are more than six score thousand persons that cannot discern between

272

---

their right hand (constructive thinking and living) and their left hand (destructive thinking and living), and also much cattle. With their many possessions?"

The writer leaves us with the personal desire of Jonah rebuked, while he closes his book with God's words of tender mercy.

Return to the Top Return to the Top

MICAH. . - p. 273

Micah means "humble."

There is a true and a false humility. True humility is not to think meanly of yourself, but, not to think of yourself at all. "Of myself I can do nothing; It is the Father that dwelleth in me that doeth the works." That is true humility.

Micah was a native of Moresheth-gath, East Eleutherpolis. He follows the three previous prophets and Isaiah (who survived him), reiterating their warnings. He died in the days of Hezekiah. He is referred to as a prophet by Jeremiah 26:18-19; his language is quoted by Zephaniah, 3:19, and Ezekiel 22:27, and by Jesus --- Matt. 10:35. He prophesied during the reigns of Jotham, Ahaz, and Hezekiah. Hebrew tradition asserts that he transmitted from Isaiah, to Joel, Nahum and Habakkuk, the mysteries of the Kabbala. One prophecy, 3:12, belongs to Hezekiah's reign, and probably preceded the Great Passover. Jer. 26:18.

273

Micah rebukes Israel and Judah for idolatries and promises restoration. A prophet is a watchman; foretelling seemed to be the least of his duties.

Mic. 2: 12   "I will assemble, O Jacob, all of thee I will surely gather the remnant of Israel." Remnant and seed are the same. A remnant does not mean a few souls, but that in every soul which is worth saving, and it will be saved.

Mic. 4:

Mic. 4: 1   Gives enlargement, peace and victory to Zion; this is the victory of the spiritual consciousness.

"The mountain of the Lord shall be established in the top of the mountains." When John of Patmos began to write the visions in Revelation, he said he was led by the Spirit into a mountain. A high place. We know why he could have these wonderful visions of what is called "the latter day." It is also called "the day of the Lord," and the "last day," but all have the same significance. "In that day the Lord shall do great things." And Israel was called upon to rejoice and be glad "in that day."

It is the establishing of the perfect consciousness.

Mic. 4: 2   "And many nations shall come, and say, come, and let us go up to the mountain of the Lord, and to the house of the God of Jacob and he will teach us of his ways, and we will walk in his paths; for

274

---

the law shall go forth of Zion, and the word of the Lord from Jerusalem."

There is no direct prophecy of Jesus, the Christ. Many so-called prophecies of him did not point to the personal Jesus, but to the Christ Spirit, the Anointed One.

Gives a direct prophecy of the birthplace of the Messiah --- the clear eyes of the prophet saw the Nazarites as the holiest, the most spiritual of the Jews, and very naturally from this sect would spring the highest expression of purity and goodness, the supreme qualifications of the Messiah.   Mic. 5: 2

Matt. 2: 56

The Old Testament is a study of race development, given by the Jews, but referring to all. From the beginning it is Christ rising from the sepulchre, whatever limits or binds. What cloth Jehovah require of thee?   John 7: 42

Mic. 7: 5

Very similar. We apply it in another way today we have no enemies except those we entertain in our own mentality.   Mic. 6: 6-8

Mic. 7:6 and

Show God's mercy the promise is complete that sin shall be utterly cast out. The sin is the sinner, not the individual. It is a mental attitude that is the sinner. "He retaineth not his anger forever, but he delighteth in mercy; He will cast all their sins into the depths of the sea."   Matt. 10: 36

Mic. 7:18-20

Return to the Top   Return to the Top

NAHUM. . - p. 275

Nahum means "comforter," also "penitent."

275

Nahum was a native of Elkosh, whose site is unknown, but traditionally was a little village in Galilee.

Believed to have prophesied after captivity of the ten tribes and between the two invasions of Sennacherib, whom Hezekiah had bribed with the treasure of the temple. The book is considered a sequel to that of Jonah.

Nahum repeats denunciations. There are three distinct predictions:

1. Sudden destruction of Sennacherib's army (Nahum 1:12) and his death in the house of his god (Nahum 1:14).

2. Inevitable capture of Nineveh by sudden eruption of river in midst of siege. (Nahum 2:6.)

3. Its utter desolation, 3rd chapter. In Nahum's time Nineveh was the largest and most opulent city in the world. It was captured by Cyaxares B. C. 606 and utterly destroyed, so that its very site was unknown a century after its fall. Nahum was considered to be the son of an Israelite captive and the vivid picture of Nineveh to have been drawn from his personal observation.

The real prophecy of Nahum is preceded by a psalm. 1:2, 2:1-3.

Nah. 1: 15	Already the bearer of glad tidings is speaking over the hills to Judah.; the final restoration of Jehovah's land and people is at hand.
Nah. 2-2	276

Prophecy concerning Nineveh, its utter ruin portrayed. And why has this fate overtaken Nineveh? Because of her lack of knowledge. The prophet tells of the ruin of the city in wonderful images.

Nah. 2: 1-3

Nah. 3: 19

Return to the Top   Return to the Top

HABBAKKUK. . - p. 277

Habakkuk, "he that embraces." We hear the Truth, but that is only half of it, we must embrace it.

Habakkuk was contemporary with Jeremiah and prophesied in Judah during the first half of the reign of Jeholakim. The book is quoted in Acts 13:41, Rom. 1:17, Gal. 3:11, Heb. 10:37. The book is in two divisions, the first a dialogue between God and the prophet; the second a sublime hymn.

The subscription " To the chief singer on my stringed instruments," shows that it was used as a psalm in which the prophet took a part, and was incorporated into the Temple service, hence he must have been a Levite. He complains of the punishment of the Jews by the Chaldeans, and is assured by God that the promises shall be fulfilled.

"Thou art of purer eyes than to behold evil and canst not look upon iniquity." This is a
literal statement. Can light look on darkness? Light could not behold darkness.   Hab. 1: 13

It does not say that the earth shall be filled   Hab. 2: 14

277

with the glory of the Lord; there is a greater assurance than that. It says that the earth shall be filled with the knowledge of the glory of the Lord. We are beginning to know it now, and that glory will absorb everything into itself.

Hab. 2: 20

Hab. 3:17-18

"The Lord is in His holy temple." We read in the New Testament, "Know ye not that your body is the temple of God?" If your body feels or looks inharmonious, say, "It is a part of that holy temple"; and insist upon it. "Let all the earth keep silent before Him." Let every belief and opinion keep silent before this truth.

He rejoices in the one kingdom, whole and entire. If we are happy because of any external thing, that thing will sooner or later fail. If we rejoice in the inner kingdom, though everything outer may fail, we shall still rejoice. "Though He slay me, yet will I trust Him." Nothing is slain but personal opinion and belief, and that is being slain all the time.

Return to the Top  Return to the Top

ZEPHANIAH. . - p. 278

Zephaniah, "the Lord is my secret." He was connected with king Hezekiah, prophesied at the beginning of Josiah's reign, B. C. 642-611. For fifty years prophecy was silent.

The book begins with the nation's sins and tells of fearful retributions,

278

The prophets all seem to have gone over and over the same warnings.

We do not have warnings today, but we suggest the necessity of giving up the things of the external as cause, of turning to the inner as the Source and Cause, and we go over and over this message. We do have to give up the mistakes we make, and until Israel did it the prophets kept reiterating over and over their message. Some of them would accomplish a great deal, and then Israel would backslide again, and perhaps it seemed discouraging to them.

Zephaniah denounces Judah but promises restoration.

Restoration and security of Israel.

"Jehovah hath taken away thy judgments, he hath cast out thine enemy the King of Israel, even the Lord,.is in the midst of thee; thou shalt not see evil any more." This is not a promise, it is a fact. If the Lord is in the midst of us, we cannot see evil or ignorance, as a reality. It is for this realization that we work.

Zeph. 3: 13-15, 17, 20

Zeph. 3: 15

Return to the Top 　 Return to the Top

HAGGAI. . - p. 279

Haggai, "Feast."

We know that Truth is a continual feast of the soul. This prophet was born at Babylon and accompanied Zerubbabel to Jerusalem. He was the first to declare God's law to the Jews after

279

their return from captivity under Darius. He urges the continuation of the building of the temple and says that the glory of the latter house shall be greater than the former.

Hag. 2: 7　Is supposed to be a reference to the coming of the Messiah. It was a prophecy of the coming of the Christ. "The desire of all nations shall come." A broad, impersonal statement. Every nation is desiring wisdom and is seeking it, each in its own way, even if it is on a low plane.

We cannot seek good without seeking God, though we may not know it.

When the prophets foretell an event externally it seems wonderful, but most of their foretelling is deeper than that. They were studying the people, their wants and desires, and they saw what kept the fulfillment of these things from them. It was always their own sins, of course.

The prophets merely interpreted the Law.

They saw that a soul would come some day who would know the Truth and would demonstrate it. Jesus was the first, but will not be the last. There have been many Christs or demonstrators of the law. It is wonderful to tell what is going to take place in the future, but we, today, are having a clearer consciousness, for we are seeing that the fulfillment is right here and now

280

Return to the Top  Return to the Top

ZECHARIAH. . - p. 281

Zechariah, "Memory of the Lord."

If all Israel had risen to that consciousness there would have been no more backsliding. Not remembering the past, but keeping the mind set on God; always remembering the law, the Lord, the divinity of her own nature would have been revealed.

Zechariah was the son of Berechiah, grandson of Iddo, tribe of Levi, and was born in Babylon. He came to Jerusalem with Zerubbabel, began to prophesy two months after Haggai, and continued two years. He was one of the captives who returned to Jerusalem after the decree of Cyrus. He prophesied through symbols and saw visions.

"Be ye not as your fathers --- your fathers, where are they? And the prophets, do they live forever?"

Restoration of Jerusalem. "I will be their God in truth and righteousness."

	Zech. 1: 4-5
The Christ Spirit predicted; "His dominion shall be from sea to sea, and from the river even to the ends of the earth."	Zech. 8: 3, 8, 16
Enemies destroyed. Rom. 9:26, Rev. 21:3.	
The "fountain opened" is the well of water within each soul.	Zech. 9: 9-10
	Zech. 12: 8-9
Supposed to refer to Jesus' second coming; as telling the exact place where His feet shall rest. Feet symbolizes understanding. "The feet shall	Zech. 13: 1
281	Zech. 14: 4

rest upon the Mount." East, turned toward the rising sun, the coming of light.

Every soul in the "light of that day," shall find his feet set upon the mountain.

	"It will neither be dark nor light, but " "one day."
Zech. 14: 6-7	
Zech. 14: 8-9	"And the Lord shall be king over all the earth; in that day shall there be one Lord, and His name one." "And Jehovah shall be king over all the earth; in that day shall Jehovah be one, and His name one." A. R. V.

One Lord and His name One.

MALACHI.. - p. 282

Malachi, "Angel or Messengers."

Malachi was the last one of the prophets, and was contemporary with Nehemiah. He prophesied B. C. 420. It is believed that he lived after the death of Ezra and the second immigration of captives, since the abuses noted in the book are exactly those which Nehemiah reformed. His book closes the writings until the new testimony of the Messiah and his work is recorded.

Malachi reproves the profanity of the priests and foretells the sudden appearance of the Messiah to purify the temple and its congregation. His writings are quoted as Scripture in the New Testament in Mark 1:2 and 9:11, Luke 1:17, Rom. 9:13. He reproves the ingratitude of Israel.

282

Impious priests rebuked. "Have we not all one Father, hath not one God created us?"	Mal. 2: 10.
Refers to office that John the Baptist fulfilled in Matt. 11:10, Mark 1:2. Anything that becomes a messenger to our souls of the Christ within is this fulfillment. Matt. 3:2, 3, 6, 10 all .refer to John the Baptist.	Mal. 3: 1
	Mal. 4:1
"All that do wickedly" the wicked deed is to be destroyed.	Mal. 4: 2
"But unto you that fear (reverence) my name shall the Sun of righteousness arise with healing in his wings."	Mal. 4: 5
Compare 11:14, 17:11, 12, 13. All refer to John the Baptist.	Luke 1:17
The spirit of Elias' message was represented by John the Baptist.	John 1: 21

283

CHAPTER 13.

NEW TESTAMENT HISTORY. . - p. 284

New Testament History	Contemporary Events
B. C.	B. C.
Birth of Jesus.	Augustus, Emperor of Rome - 27
	Death of Herod. - 4
	A. D
	Cyrenius (Quirinius), Governor of Syria (2nd time) - 6
	Revolt of Judas of Galilee
	Tiberius, colleague of Augustus 12
	Death of Augustus 14
	Caiaphas, High Priest 25
	Pontius Pilate, Procurator - 26
A. D.	
8 - First visit to the Temple.	
27 - Baptism of Jesus.	
30 - Crucifixion and resurrection of Jesus.	
Descent of the Holy Spirit at Pentecost.	
35 - Martyrdom of Stephen.	

Conversion of Paul(?)	
	Caligula, Emperor - 37
	Herod Agrippa I, King of Judea and Samaria
38 - Paul's first visit to Jerusalem.	Claudius, Emperor - 41
	Death of Herod Agrippa I - 44
44 - Martyrdom of the Apostle James. Paul's second visit to Jerusalem. Epistle of James (?).	

284

New Testament History A. D.	Contemporary Events
45 Paul's first missionary journey; Antioch to Cyprus, Pisidia, Derbe, Lystra, etc., and back.	
50 Jerusalem council. Paul's third visit.	
51 Paul's second missionary journey; Antioch to Cilicia, Lycaonia, Galatia, Troas, Philippi, Thessalonica, Beraea, Athens, Corinth.	
52 Paul at Corinth,	Claudius

remaining one and a half years.	banishes Jews from Rome 52
53 I and II Thessalonians.	Antonius Felix, Procurator 53
54 Paul's fourth visit to Jerusalem. Short stay at Corinth.	
Third missionary journey commenced.	Nero, Emperor 54
Paul at Ephesus two years.	
	Revolt of Sicarii (Acts 21:38) 55
57 I Corinthians (from Ephesus); II Corinthians (from Macedonia).	
58 Galatians and Romans (from Corinth) Paul's fifth visit to Jerusalem. Imprisonment at Caesarea, lasting two years.	
60 Paul before Festus, and sent to Rome. Two years a prisoner there in hired house.	Porcius Festus, Procurator 60

285

New Testament History	Contemporary Events
A. D.	A. D.
62 Epistles (letters) to Philippians, Ephesians, Colossians, and Philemon (from Rome).	Josephus at Rome 62
63 Paul released (?).	
64 Paul visits Crete and Fire of Rome.	

	64
Macedonia. I Timothy and Titus.	Nero's persecution
66 II Timothy (from Rome)	
Martyrdom of Paul and Peter (?) About this time the Synoptic Gospel is written, also perhaps Jude and I and II Peter.	Beginning of war be tween the Jews and Rome 66
	Vespasian, General, in Palestine 67
68 John's Apocalypse (Revelation). Epistle to Hebrews (?).	Galba, Emperor. 68
	Otho, Vetellius, and Vespasian, Emperors. 69
	Destruction of Jerusalem 70
	Titus, Emperor 79
	Destruction of Pompeii and Herculaneum . . .
	Domitian, Emperor . 91
	Domitian's persecution 95
96 John's Gospel (?). Epistles and	Nerva, Emperor 96
98 Death of John (?).	Trojan, Emperor 98

286

Return to the Top   Return to the Top

HEALING WORKS OF JESUS.. - p. 287

Works	Matt.	Mark	Luke	John	Place	Method
Man with unclean spirit		1:23	4:35		Capernaum	Word
Peter's wife's mother	8:15	1:30	4:38		Bethsaida	Touch and word
Multitudes	8:16	1:32	4:40		Capernaum	Touch and word
Many demons		1:39			Galilee	Exorcism
The leper	8:2	1:40	5:12		Gennesaret	Word and touch
Many sick of palsy	9:2	2:5	5:20		Capernaum	Word
Man with withered hand	12:9	3:1	6:6		Capernaum	Obedience, touch
Multitudes	12:15	3:10			Gennesaret	Exorcism
Garasene	8:28	5:1	8:26		Gadara	Word
Jarius' daughter	9:18	5:22	8:41		Capernaum	Word, touch
Woman with issue	9:20	5:25	8:43		Gennesaret	Woman's faith
A few sick folk because of their unbelief	13:58	6:5			Galilee	Touch
Multitudes	14:34	6:55			Gennesaret	Touch of garment
Syrophenician's daughter	15:22	7:25			Tyre	Woman's faith
Man deaf and dumb		7:32			Decapolis	Word, touch
Blind man at Bethsaida		8:22			Bethsaida	Word, touch
						Word,

Works	Matt.	Mark'	Luke	John	Place	Method
Lunatic child	17:14	9:14	9:38		Tabor?	touch, disciples' failure
Blind Bartimaeus	20:30	10:46	18:35		Jericho	Word --- man's faith
Centurion's servant	8:5		7:2		Capernaum	Word
Two Blind men. See that no man know it	9:27				Capernaum	Touch, word
Dumb Daemoniac.	9:32				Capernaum	Exorcism
Dumb Daemoniac.	12:22		11:14		Capernaum	Exorcism
Multitudes	4:23		6:17		Galilee	Teaching and healing
Multitudes	9:35				Galilee	Teaching and healing

10

287

Works	Matt.	Mark'	Luke	John	Place	Method
Multitudes	11:4		7:21		Capernaum	Proofs, John
Multitudes	11:4		9:11	6:2	Bethsaida	Word
Great Multitudes	15:30				Decapolis	Word
Great Multitudes	19:2				Judaea	Word
Blind and lame in temple	21:14				Jerusalem	Word
Widow's son of Nain			7:11		Nain	Word
Mary Magdalene and others			8:2			Word
Crooked			13:10		Jerusalem	Word,

woman			touch
Man with dropsy	14:1	Jerusalem	Word, touch
Ten lepers	17:11	Samaria	Word
Malchus	22:51	Gethsemane	Touch
Great Multitudes	5:15	Gennesaret	Word
Nobleman's son.	4:46	Cana	Word
Impotent man at Bethesda	5:2	Jerusalem	Word
Man born blind	9:1	Jerusalem	Word, touch
Lazarus	11:1	Bethany	Word

None of the cases of healing are recorded by all of the four Evangelists. Five are given by John, all special instances but one.

Casting out devils by non-disciple recorded in Mark 9:39, Luke 9:49.

Jesus turns water into wine --- Cana , John 2:1-11.

Draught of fishes --- sea of Galilee , Luke 5:1-11. Calms the tempest, sea of Galilee; Matt. 8:23-27.

Feeds five thousand --- Decapolis , Matt. 14:15-21.

Feeds four thousand --- Decapolis , Matt. 15:32-39.

Causes great draught of fishes --- sea of Galilee; John 21:1-14. Last miracle.

289

Return to the Top   Return to the Top

CHAPTER 14. - THE FOUR EVANGELISTS.

Part I. Evangelical History before the public ministry of Jesus;

PART I.

Three Jewish parties: Pharisees, orthodox teachers of Mosaic Law Sadducees, worldly minded priests, aristocrats Essenes, ascetics, lived apart from social and civil life in communities their lives were simple, devout, orderly they did not indulge in the fierce pride of the Pharisees or the scornful skepticism of the Sadducees.

Evangelical History before the public ministry of Jesus is supposed to cover the space of thirty years and six months.

On comparing the first three Gospels, the Synoptic Gospels, we observe in them great similarity both in substance and general arrangement. The outline of the course of events given in these Gospels is almost the same, from the ministry of. John the Baptist onwards.• The common character is very accentuated when we compare them with John, the contents of which are widely different.

"In the beginning was the Word, and the Word was with God, and the Word was God."
The same, God, Divine Mind, that originates, that forms, that upholds, that sustains by   John 1: 1-4
being present with, was in the beginning. "All things were

289

---

made by Him, and without Him was not anything made that was made."

Luke 1: 1-4	Luke's Preface.
John 1: 1-19	John's Preface.
Luke 1: 5-25 and 57-79	Conception and birth of John the Baptist.
Luke 1:26-38	Revelation to Mary.
Matt. 1:18-25	An angel appears to Joseph. Only Matthew relates this incident. Virgin, Hebrew word for young woman, married or unmarried. Joel 1:8.
Luke 1:39-56	Mary visits Elizabeth.
Matt 1: 25	The birth of Jesus.
Luke 2: 1-7	The birth of Jesus. Luke relates more in detail concerning the conception. An angel spoke to Mary it was a message to her soul and those who wrote this could not have understood that message.

There is something higher than the mystery of Jesus' birth. Imagine some one today becoming conscious that all life is Divine. Mary seems to be the first to whom this Divine revelation came.

It came within her own consciousness that the Holy Ghost, Divine illumination, should come to her.

The first necessity of all creation is light, the first creative word was "Let there be light."

Every child is a creation, a Divine creation.

Divine illumination came to Mary's inmost consciousness first the light, understanding, inward

290

illumination, then next " the power of the Highest shall overshadow thee."

The world's conception of birth is completely overshadowed by the new conception that reveals the higher birth the highest vision shall overshadow every human conception.

Immaculate, pure, without blemish all conception is immaculate because conception, growth and unfoldment are by natural law which is Divine Law.

The birth of Jesus was a natural birth and Mary was the first who perceived the divinity of it.

It seems that the other children that Mary brought forth were not of such a high consciousness as Jesus. Perhaps she did not rise quite to that height again. Jesus was surely the offspring of her purest consciousness.

Jesus had no advantage over us from God but from his mother. Mary was a consecrated soul. The divinity is what is born of God and that means every child.

Equally as wonderful were the conceptions of Elizabeth and Sarah. They were both supposed to be beyond the age when woman brings forth. According to human law, Elizabeth could not have brought forth John the Baptist, but there are laws of birth which are beyond our comprehension today. Why were not the products of Elizabeth and Sarah as high as the product of Mary?

291

Note that the message in Sarah's case came, to Abraham, and Sarah doubted it in the case of Elizabeth, the message came to Zachariah, who carried it to her.

Promises are predictions; not something granted by an arbitrary will as a reward. This is the childish notion, but in maturer consciousness it is revealed that all is law, and by law. Predictions concerning the Messenger that shall prepare the way for the "coming" of the Lord. After the preparation, the Lord comes quickly. What is the preparation? It is called the cleansing. Of what? Of mentality, the dissolving of personal opinions and beliefs. These alone obscure our clear vision of Truth. The "Voice" in the "wilderness" always sounds. Preceding every thought and word and deed is the guiding consciousness that directs. The "wilderness" where all seems bare is filled with this "Voice."

The parents of John were righteous, "walking in all the commandments"; yet Zacharias doubted the "Voice" that spoke or predicted beyond human thought of possibilities. For this, personal expression was stilled until the revelation was received and accepted. If we will accept it, we may receive far beyond what human belief decrees.

There is no prayer of faith, until we seek beyond human understanding!

292

Return to the Top   Return to the Top

Elijah was the Precursor of John - p. 293

Mal. 4:5	Mark 9:11-13
Matt. 11:14	Luke 9:8
Matt. 16:13-16	Luke 1:17
Matt. 17:10-13	John 1:19-23

To go forth in the spirit and power of any individual is a lesser consciousness than to go in the spirit and power of your own Christ.

In the case of Mary, the message came to her own soul, not through another. Self-revelation is the highest nothing can be as high as that which comes to one's own soul. We must see the Truth with our own eyes. We may read or hear what another says, but we must weigh it and either receive or reject it for ourselves.

Jesus disclaimed human origin, either through father or mother, and claimed his divinity and claimed the same for each of us. John 13:3 and John 16:28. It is he who taught us to say, "Our Father," and he said, "I ascend unto my Father and your Father, unto my God and your God." John 20:17. Even more emphatically he says, "Call no man on earth your father, for One is your Father, which is in heaven." Matt. 23:9. He could have said, "Call no woman on earth your mother, call no flesh your Origin, but Spirit." We are heirs of God, equal heirs with Jesus, the Christ. This is the promise of the

293

Christ consciousness to all men. Birth not of the flesh but of Spirit, of a virgin mind that knows not man, man's law of intellect. "Born not of blood, nor of the will of the flesh, nor of the will of man, but of God." John 1:13. My origin is yours, declares this Divine man. I am the elder brother of this great family, "first born" because first to perceive man's true origin.

	The genealogies of Jesus. These accounts differ. Luke goes back from Jesus to Adam;
Matt. 1:1-17	Matthew goes back from Jesus to Abraham, and uses the word "begat," true reading,
	"Joseph begat Jesus who is the Christ." Both authors in giving the genealogy of Jesus trace
Luke 3:23-38	him back through Joseph, not through Mary.

Matt. 13: 55

Luke 2: 48

Luke 3: 23   > All refer to Jesus as son of
                                    Joseph.

John 1: 45

John 6: 42

It is believed that Mark wrote the earliest Gospel, yet it gives no account of virgin birth or birth at all.

	Paul makes very clear his thought of Jesus' birth, "born of woman under the law." God's
Gal. 4: 4	law is the law of perfect expression.

The festival of the immaculate conception of Jesus is traceable in the Greek church from the fifth century, and in the Latin dates from the seventh century.

John Duns Scotus, in 1307, introduced the

294

---

teaching releasing Mary from the Jewish idea of original sin. They did this to free her from the taint of Adam's sin; her ancestors were left in their original condition.

Pope Pius IX issued a decree at Rome on December 8, 1854, declaring the doctrine to be an article of Catholic belief; the decree is accepted throughout the Roman church.

Presentation of Jesus in the temple. "And Joseph and his mother marveled at those things which were spoken of him." Luke 2: 22, 33, 38

The wise men directed to the infant Jesus with gifts. Every babe has received all the gifts of wisdom. The wise men worshiped the child, not the mother; true wisdom does not worship any personal Jesus, but the Spirit, God. Matt. 2: 1, 12, 23

Herod's cruelty causes them to hurry into Egypt. Luke 2: 40-52

Jesus at the age of twelve, time when Jewish child became the son of the law. Heb. 5: 8

Jesus learned obedience. Matt 3: 1-12

John the Baptist and his ministry. John came in the spirit of Elijah, with the same vigorous arraignment. Mark 1: 1-8

Luke 1: 80

He replies definitely, "I am not the Christ. I am not Elijah." Luke 3: 1-18

Jesus says, "Among those that are born of Jno 1: 15-28

10a

Jno. 1: 20, 21, 25

295

Luke 7: 28

---

women (believe this to be the truth of their birth) there is not a greater prophet than John the Baptist but he that is least in the kingdom of God (he that has the very least consciousness of the presence of God) is greater than he." The difference between John the Baptist and Jesus, John said, "I am not the Christ," and Jesus said, "I am the Christ," difference in realization of being.

Return to the Top  Return to the Top

---

Part II. Events of about six months from Jesus' baptism, until the beginning of the ensuing Passover.. - p. 296

PART II.

Events of about six months from Jesus' baptism, until the beginning of the ensuing Passover.

BAPTISM OF JESUS:

Matt. 3:13-17	Luke 12:50
Mark 1:9-11	Matt. 20:23
Luke 3:21-22	Mark 10:38

John 4:2. Jesus baptized not.

Coll. 2:11-12. Inward baptism.

1 Cor. 1:17 Jesus the Christ sent to preach the gospel, not to baptize.

Rom. 6:3-4. Baptism.

John 1: 35-36, 43	Baptism, inward illumination. "Day after" baptism John said, "Behold the Lamb of God." Second day Jesus calls disciple.
Phil. 2: 1-2	Third day after was the marriage in Cana.
	Three days, purifying process Jesus was beginning in his unfoldment,

296

John the Baptist's method of baptism is to pass:

Matt. 3:11	John 1:30-33
Mark 1:7-8	John 3:30
Luke 3:16-17	Acts 1:5

Acts 19:1-6 and 18:25

THE TEMPTATION OF JESUS:

Matt. 4:1-11	Luke 4:1-13
Mark 1:12-13	

James 1:13. Let no man say when he is tempted, "I am tempted of God", for God cannot be tempted with evil doing, neither tempteth he any man. But every man is tempted when he is drawn away of his own lust and enticed.

Devil, father of lies, liar from beginning, a superstition no reality. Jesus did not kill Satan, nothing to kill, no reality. Give no place nor power to the devil.

Personal desire tempted him. Three classes of temptation, supply, personal protection, personal dominion.

Fasting, the attitude of meditation and concentration upon Truth which frees us from personal opinions and personal beliefs. Jesus worked out the problem of personality. Jesus fasted forty days. Forty is one of the numbers in the Bible which we can take to mean a finished condition, or the completion of a certain process in development,

297

---

the working out of a problem. Moses was on the Mount forty days, communing with God, getting the law.

Ex. 24: 18 — In the law of Moses it is said of the wicked man that he must be given forty stripes, no more. It means complete redemption from that condition. The Israelites wandered forty years in the wilderness, until all the generations that had done evil in the sight of the Lord had been consumed, until that condition had been finished.

Saul and David and Solomon each reigned forty years.

Paul says, "Of the Jews five times received I forty stripes save one." II Cor. 11:24.

Jesus was seen forty days after his resurrection before his ascension. Acts 1:3.

In these instances, and many more, we plainly see the significance of the number. Jesus had a problem to work out in the forty days.

Angels ministered to him, waited upon him, during his fast of forty days.

Mark 1: 13 — Jesus gained the realization, I am the son of God.

"Thou shalt not Tempt the Lord thy God." Deut. 6:16.

Deut. 6: 13,
10: 20

Jos. 24: 14 — >Thou shalt worship the
                              Lord thy God.

1 Sam. 7: 3

Jesus had not conquered every sense of separation,

298

---

but he did not yield to it when he commanded Satan to go behind him. Personality tempts us to believe that we can do all things, and promises us all the kingdoms of the world (showing that these belong to the personal belief), if we will but acknowledge personal power. To be nothing apart from God is not so easy after the high vision of man's sonship.

Return to the Top   Return to the Top

PART III.

Some of the events that took place during the twelve months from the beginning of the first Passover.

Jesus visits Capernaum.

Jesus goes to Jerusalem at the Passover, and casts the traders out of the temple.

John 2: 12

"Destroy this temple, and in three days I will raise it up." Moses erected the tabernacle, the temple of the Israelites in their journeyings. Solomon reproduced it in an edifice of wood and stone upon its rocky foundation. Jesus raised his body, the temple of the living God, in three days through his consciousness of its perfection. By resurrection he proved that "death" has no power over the body.

John 2:13-25

John 2: 19

Heb. 2:14-15

By his ascension, when that body disappeared from human gaze, he proved that here and now this mortal is to put on immortality.

299

2 Cor. 5: 4

Mortality is "swallowed up of life" when consciousness of Life fills our vision. Transformation of the body is not a change in the real nature of the body, but a change in our thought of the body. Soul and body are one "what God hath joined together, let no man put asunder".

In the parable of the marriage feast, we find the lesson of the necessity of the perfect body accompanying the right understanding of the unity of individual soul with Spirit. Matt. 22:2-14. The invitation to the wedding is the appeal of Truth to each individual to come into recognition of its oneness with Universal Spirit. A guest appears without a "wedding garment", and he is ejected from the banquet. Clothing is the external and symbolizes the body. The wedding garment is the perfect body.

As the guest not having on this garment shames the master of the feast, so does the soul that professes conscious unity with Spirit dishonor Spirit if it is not "clothed" with perfect body.

We may claim this unity, but "by their fruits ye shall know them". If the visible does not manifest Divine likeness, it simply proves that our souls are wedded to some thought besides that of Spirit. If our soul is consciously wedded to Spirit, all outward conditions will be perfect.

In the barren fig tree we find the same suggestion.

Matt. 21: 19

300

The tree was incomplete beautiful in appearance, yet it was cursed (rejected) because it bore no fruit, brought forth no proof of its source.

"For He is our peace, who hath made both one" (Spirit and flesh --- inner and outer),." and hath broken down the middle wall of partition" (rent the veil of ignorance). "Having abolished in his flesh the enmity, for to make in himself of twain" (that which had been supposed to be two, differing in kind and substance) "one new man, so making peace.

Eph.2:14-15

John 10: 9

"By me (by my understanding) "if any man enter in" (realizes unity) "he shall be saved, and shall go in and out and find pasture." Such consciousness shall find eternal substance (pasture) in both inner and outer, in both soul and body no belief in separation; for all things are seen in oneness with Spirit.

Reconciliation, or the reconciling of our thoughts to the consciousness of agreement everywhere between all things, is at-one-ment. Jesus did not make the atonement or at-one-ment, He revealed it. At-one-ment is an eternal verity. Truth is not man-made, but man-discovered.

Col. 1: 20

Eph. 1: 10

In the light of the at-one-ment, that which was supposed to be "twain" is seen as one; the veil of separation is "done away" between God and God's expression and manifestation, between the universal man as Spirit, and the individual man

301

---

as living soul and body. The inner and outer are now known as one; the Spirit and the body are now known as one and the sacredness of the body begins to be realized.

1 John 4: 2

"There is but one temple in the universe, and that is the body of man. Nothing is holier than that high form." Oneness with God. is the at-one-ment that redeems each living soul and body, by revealing the Truth that living soul and body are one in the Spirit that begot them. Realization of unity is at-one-ment that each must make as Jesus did, by seeing and knowing man's oneness with God.

John 3: 1-21

Jesus' discourse with Nicodemus, a ruler of the Jews. The great teacher shows very plainly the necessity of man's birth into the consciousness of his divinity and his divine possibilities into the understanding that that which is born of Spirit is Spirit, like begets like. Man cannot enter the kingdom of God consciously until he realizes the truth of his unity with God, the omnipresence of God.

Jesus was the "first born among many brethren," the first to know the truth of himself as the son of God.

The Christ principle (the individual soul) is the Son and not a Son.

Mark 12: 35-37

Jesus tarries in Judaea where his disciples baptized. John the Baptist asserts the greater

John 3: 22-36

302

---

spiritual unfoldment of Jesus, "He must increase, but I must decrease."

Passing through Samaria, Jesus meets the woman and did not prevent her carrying the deep teaching he gave her to her people. To worship

John 4: 1-42

is to acknowledge the Truth that God is Spirit.

John 4: 20, 21, 23, 24

True worship is to acknowledge in Truth.

Matt. 4: 12

After John the Baptist's imprisonment, Jesus re tires into Galilee and exercises his public minis try. He heals the son of King Herod's officer, who lay sick at Capernaum.

Mark 1: 14-15

Jesus goes to Nazareth, where he preserves his own life by his consciousness of the presence of God. Later he fixes his dwelling at Capernaum.

Luke 4: 14

John 4:43-54

The call of Simon, Andrew and James and John.

Matt. 4:13-16

Jesus attended by some of his disciples teaches and heals throughout Galilee.

Luke 4:15-31

Matt. 4:18-22

Don't get devoured by
the costs
Get an accurate picture
of New Thought Education
DivinitySchool.net

Mark 1:16-20

Luke 5: 1-11

Matt. 8: 14

Matt.14: 23-25

Mark 1: 29-39

Luke 4:38-44

PART IV.

Some of the events that took place during the twelve months, from the beginning of the second Passover:

Matt. 5th, 6th and 7th chapters.

The Sermon on the Mount. Read Matt. 4:23-25 --- Jesus' Philosophy of Life.

"Seeing the multitudes, he went up into a

303

___

mountain, and when he was set (settled, established in consciousness) his disciples came unto him."

We can lead others to truth only as we ourselves are fully convinced, and are faithful to what we know.

"He opened his mouth and taught" --- and the first words are benedictions. Jesus says first of all that conditions that men usually pity are blessed! They that are poor in spirit, that mourn, that are meek, that hunger and thirst after righteousness, all of these conditions are called happy. Why? Because they that have riches of intellect, of externals, of abundance in the outer, are likely to feel satisfied with that which is temporary. While those who feel empty will seek. Those who have little to boast of in the personal, will turn to the eternal and greater things of Truth.

Then Jesus says, Those who are merciful shall be treated with mercy, or as he said later --- The measure we give forth shall return to us --- The judgment we pass on others shall be passed on us. Jesus simply speaks of a law --- the measure we mete, the judgment we make is in accord with the development of our thoughts, it proves the level of belief upon which we are living and no better than this can be given out or received by us.

The pure in heart shall see God," because that is the God-like consciousness.

304

___

The peace-makers are called happy. While they are exercised in bringing peace to others, they are, by the law we have spoken of, making their own peace, for they are in the realization of peace.

Two more strange "Blesseds" fall from his lips. They that are persecuted for righteousness' sake, and they that are reviled and spoken evil ag6inst falsely --- not the attitude of a martyr, but the love of Truth for Truth's sake is encouraged here.

"Rejoice and be exceeding glad, for great is your reward in heaven," harmony or consciousness. It is the inward and not the outward that counts. It is such souls that give to living its positiveness that preserve its interest such souls are a light to the world, because [they are] from personality. They can shine and let their light be seen since all the glory of it will be ascribed to God.

Matt. 5: 20

Matt. 5: 17

Truth comes not to destroy anything real, but to fill our recognition to the full of that which eternally is.

Men of old spoke according to their light. Jesus speaks from the fuller light. "Be ye perfect."

Matt. 5:21-48   Matt. 5:38-39, 42
Matt. 5:23-25   Matt. 5:43-48

305

Return to the Top   Return to the Top

THE LORD'S PRAYER. - p. 306

"1. Our Father which art in heaven." Our Source is our Father, or Feeder, that which sustains all life is Infinite, it is Divine. "In heaven," in my spiritual consciousness.

Thus I am given to eat, to be fed or sustained by the Tree of Life, in the midst of the paradise of God.

"I Am the tree of Life," for Christ says, "I am the life, and my environment is the heaven or paradise of God.

My Source, my Life, which sustains and strengthens my thought and body every day, is the I am that I am as perceived in spiritual consciousness.

2. "Hallowed is thy name."

Thy name is Life, and wherever thy word is spoken, thy perfect Life is manifest.

Thy name of Life is holy, it is wholeness, it is unity, whether invisible or visible. Infinite Truth holds all Life in oneness. Thy name, nature, whether in God or man, is hallowed by that consciousness which understands. "I will give him a white stone and in the stone a new name written, which no man knoweth saving he that receiveth it."

"Thou hast not denied my name, I will write

306

upon you the name of my God. I will confess your name before my Father."

3. "Thy kingdom is come, thy will is done on earth as it is in heaven."

"I will give him power over the nations to rule them with a rod of iron."

Divine Consciousness in me rules and breaks every mortal claim.

4. " Thou givest us each day our daily bread."

"I will give him to at of the hidden manna."

"I am that bread which came down from heaven of which if any man eat he shall never hunger if any man drink of the water that I shall give him, he shall never thirst, for it shall be in him a well of water springing up into everlasting life."

The "daily bread" is now the "hidden manna," or it is that substance which I now am able to draw from within myself. I am the Source, in oneness with God.

5. "Thou forgivest our trespasses as we forgive those who trespass against us."

While we see anyone as sinful, we cannot receive consciousness of our forgiveness of sin. While we see anyone else as sick, as a reality, we cannot be conscious of our wholeness.

Until we die to all beliefs of evil or ignorance, we cannot be free from such claims.

307

As soon as we come up over all that claims separation anywhere, we forgive those who have trespassed against us and are forgiven and now we see that no hurt can come near our dwelling.

"He that overcometh (forgiveth first) shall not be hurt of the second death" --- shall be forgiven all --- shall be able to put out all without "hurt."

6. " Thou leadest us not into temptation but dost deliver us from all evil:"

"I will also keep thee from the hour of temptation. I will make thee a pillar (firm consciousness) in the temple of my God, and thou shalt go no more out."

7. "For thine is the kingdom, the power and the glory forever."

By him that overcometh, personality is surrendered. The One (God) is known as all power and glory forever.

"To him that overcometh will I grant to sit with me in my throne." Are we ready to sit upon this throne? Not until we are ready to ascribe to the One that is Universal all in all --- the Christ -- Divine Truth that rules the world.

There are seven divisions in the "Lord's Prayer," as there are seven promises to the churches.

Upon giving modestly, praying likewise.

Matt. 6   308

Get not by striving; learn to acknowledge that which is. That which is can never be taken from you as may that which is acquired. Have satisfaction in Truth invisible rather than in appearances.

Matt. 6: 19-21

Matt. 6: 22-23

The single eye sees but one --- the Good.

Matt. 6: 24

Acknowledge one power.

Matt. 6: 25-33

Drop questionings. Truth is; acknowledge it, for it is fullness now. Seek to know only God and in finding God you shall find all else.

Matt. 7: 1-5

Seeing faults in others is because our vision is faulty.

Matt. 7: 6

Do not force your perception of Truth upon unwilling ears.

Matt. 7: 7-8

Ask, seek, knock --- open to; since God is omnipresent, you shall receive.

Matt. 7: 12

"Love your neighbor as yourself."

Matt. 7: 13-14

Conceptions are many; all shall go the way of destruction. Consciousness of the Truth, only goes in the way of life.

Matt. 7: 15-20

Matt. 7: 22-23

Deeds, not words, testify to true consciousness. Iniquity is separation. That thing accomplished, this belief of personal power, is not acknowledged in truth.

Matt. 7: 24-27

Founded upon the basis of Truth, all is enduring. Words and deeds are sacred expressions of the Eternal. That which is done from a personal standpoint is temporal only and will not long endure the tests of the Spirit,

309

Matt. 7: 29 — He who knows speaks authoritatively because he is aware that not he personally speaks or teaches, but the Spirit in him that is the Spirit in all.

Matt. 11:2-19
Luke 7:18-35 — Jesus' answer to the disciples sent by John with the question, "Art thou he that should come, or do we look for another?"

Matt. 11: 20-30 — Jesus upbraids the cities wherein most of his mighty works were done, because they repented not.

Luke 7:36-50 — A woman anointeth Jesus' feet. "Thy faith hath saved thee; go in peace."

Matt. 9:35 & 12: 22-37 — During Jesus' second visit in Galilee he heals a demoniac and the Scribes and Pharisees will not accept his teaching.

Mark 6: 6 & 3: 20-30 — "Verily I say unto you, all sins shall be forgiven unto the sons of men, and blasphemies

Luke 8: 1-3 & 11: 14-23 — wherewith soever they shall blaspheme. But he that shall blaspheme against the Holy Ghost (the consciousness of the whole Truth) hath never forgiveness, but is in danger of eternal damnation."

Mark 3:28-30 — The individual who rejects the light of understanding, who chooses to walk in darkness, literally damns, impedes his unfoldment into the consciousness of his own being, the truth of his divinity, of his divine possibilities by virtue of his being a child of God. We see things by our attitude. Truth is passing by us now. Are we going to see it or let another see it for us --- our choice is

310

---

to receive it or reject it. The will is the soul's power of choice. Jesus claimed his divinity.

John 6: 25-65

Is the second sermon of importance 'that Jesus gave. Read carefully.

John 6: 66

From that time many of his disciples went back and walked no more with him. They did not understand his message --the bread is the substance, the blood is the Life. The I

Mark 10: 18

Am he claimed for himself is the same I Am that you know as yourself. It was because they thought he was speaking from the personal standpoint that they were unable to understand his words. Never did this divine man refer to his personality. "Why callest thou me good, there is none good but one, God." The third of his three principal sermons is John 14, 15, 16 and 17 chapters.

John 14: 2

John 14: 3

John 14: 5

John 14: 8

"Many mansions" in the "Father's House."

In infinite consciousness there are many stages

of unfoldment, but each soul's consciousness has place in or belongs to the Infinite.

Every pure life. "goes before" and "prepares the way" by opening itself to this perfect consciousness. Thus it proves to others what each may do.

If anyone consider this a literal, visible "way"

and "place," listen to Jesus answer in verses six and seven.

If he still insists upon knowing a God apart, let him understand Jesus' words, "He that hath

311

Isa. 9: 6

seen me, hath seen the Father." Christ's Spirit is born in us, child embodiment of meaning.

John 14: 9

John 14: 12

God and man are one. Man's work or accomplishment in Truth is God's activity and fruit.

John 14:13-14

God is no respecter of persons. Jesus' personality was no more to God than yours. God has respect only to consciousness. I Cor. 12:9-11.

John 14:15 and 21

In the name Divine, not in the personal name Jesus. The Divine name is the Divine Nature; it is yours and mine as it was Jesus' name or nature.

John 14: 16-17

Eph. 1:21	Phil. 2:9-10

John 14: 18

Luke 10:20	Rev. 3:5

John 14:19-20

John 14:22-24

The Divine Nature of which each of us is a partaker asks what it will, because it already inherits all.

John 14: 25-26

Proof of love is in deeds, not in words.

The Comforter is Divine Understanding. Jesus prayed that his followers might open their vision to it.

Assurance that they shall understand him.

What he did we may do when we see him as he is, one with his Source, as we are too.

God is manifest to us, or clearly seen and felt as all Presence, only through love, conscious unity.

Love keeps the sayings of Truth. "The sheep hear my voice."

The personal voice speaks so long as the outer ear must be satisfied. When the inner hearing is opened, the voice of Spirit teaches the higher

312

meanings of the spoken word. Happy are we when we can hear every word translated into spiritual meaning.

Not as the world giveth, not uncertainly, not sparingly, not demanding return, giveth Divine Love unto us. But giving because it loves to give; because it has everything to give giving because we are its offspring. Divine Love gives fully, freely, saying, "All mine is thine." God is love.

John 14: 27

John 14: 28

"My Father is greater than I." The Universal is greater than the individual, though the individual is just like the Universal, is of its very substance, has received of its Life, Intelligence and Power.

John 14: 29: 30

John 14: 31

"The Prince of this world." The most cherished world belief, no matter in what veneration it is held, affects not the soul whose consciousness of eternal things is clear. Such a soul has nothing in it to which belief can attach itself.

John 15:1-2

John 15:4-5-6

This I commanded your fathers, that they obey my voice. Obedience is always the test of love. Words of love are proved by deeds of love. Man is the true vine, thoughts are the branches, words and deeds the fruit.

Believe in man; know that God is manifest as man, that God thinks, speaks and works as man. Abide in this consciousness and every thought (branch) shall bring forth good, or shall see good only.

313

John 15: 7	Abiding thus, thought staid upon good, ask anything, for your asking will be simply seeking what is.
John 15: 8	Know that good fruit glorifies God, its Source, rather than you. This was Jesus' consciousness, so must it be the understanding of all of Truth's followers.
John 15:9-10	
John 15: 11	Let Love (conscious unity) be the dominating impulse. It will lead us into perfect obedience.
John 15: 12	Realization of Love (conscious unity) is joy.
John 15: 13	Therefore let love to God and to fellow-man be your delight.
John 15: 14	Prove your love (conscious unity) by deeds; even life may be subjected to love.
John 15: 15	Again, prove yourself --not words but deeds count.
John 15: 16	It has been thought that humility compelled us

to think of ourselves as "servants" and was it not sufficient to be as a servant in the Lord's house? God does not so look upon us, but knows us as His children, and when fuller grown, as His sons. Jesus tells why. John 8:35, Gal. 4:6-7.

"I have chosen you and ordained you that ye should go and bring forth fruit." We are the chosen of Truth, not a few selected ones, but all of which the disciples are types. Jesus chose his twelve followers knowing their hearts and their readiness to follow Truth. Their choice was really made before Jesus called them.

314

Repetition of the all important lesson --- Love. The world thought cannot love that which is to be its destruction. When supposed ignorance loves

John 15:17-18-19

Truth, that is the end of ignorance.

John 15: 20-21

I gnorance alone persecutes.	John 15: 22
The more light, the more responsibility. There is no excuse for ignorance after an awakening of the soul to Truth.	John 15: 23 John 15: 24
Jesus is gradually coming to his highest statement, "I and my Father (Source) are one."	John 15: 25
Seeing and not believing is hatred. Believing and accepting is love. Unbelief and rejection is hate. Sins of the parents are visited upon "those that hate me," or reject the Truth because of unbelief. Explains also verse twenty-three.	John 15: 26-27 John 16:1-2-3-4
All hatred is without cause. God is the only cause. If we will remember this, it will soon allay all inharmony. Love is the only cause.	John 16: 5-6-7

The Spirit of Truth always testifies to the truth of man, this "me" we are told to believe in.

Verse one. Rom. 8:16.

The consciousness that we gain in our communions shall abide with us and be our stay, if we will remember, when the world claims press heavily.

It is always best for us to look to invisible Spirit for guidance and help than to personality or outer helps. Each soul has within it all that

315

---

it needs; the sooner it finds and trusts this, the better.

John 16:8-11	Reprove, convince. Our first conviction must be of our mistakes, then are we willing to surrender our opinions and beliefs. After this, we are
John 16: 12	willing to listen to rightness, and to form true judgment.
John 16: 13	Until the individual can receive, admit, acknowledge Truth, it is useless to force it upon

him. Give each what he is ready for, and by use of that he will be made ready for more, or for fuller revelation.

Being ready is when the Spirit is acknowledged, since it is always with us. We do not open to it, which is to acknowledge it, until we are ready. When acknowledged, we see it as our guide, and following its voice we hear truth more and more clearly until its fullness is realized. This is the assurance to all, and will be realized when we can receive it.

John 16: 14	The Spirit shall glorify man, for it knows of man.
John 16: 15	All that is God belongs to man, and each can say, "It is mine," as soon as he has seen it.
John 16: 16, 17, 19, 22	Temporal joy is fleeting, vacillating. We see at times and then we say we do not see. In consciousness we shall see always, and this is the enduring joy that no thought can take away from us.

316

Constantly does this loving teacher remind his listeners of the essential things. Ask in my name, he urges, because he knows that only the Divine Name or Nature inherits from God. Only that which comes from the Source receives of the Source.	John 16:23-24 John 16: 25

Jesus' words are figurative because, as yet, they cannot comprehend; their acceptance of the Spirit of Wisdom as their own can alone lead them into the understanding of all Truth. This clear consciousness shall be ours.	John 16: 26
	John 16: 27
This verse shows this too.	John 16: 28
	John 16: 29-31
God loves his own.	
Each comes from God and returns to God, until neither coming forth nor returning is known. This is called birth and death, but when the truth of it is clear, we shall know ourselves constantly evolving in God, of God and like God.	John 16: 32
	John 16: 33
Over-confidence in personal belief is always rebuked.	John 17:1

This prediction is a warning not to speak words that we shall not fulfill in act. The aloneness of the soul to human thought may be its closest companionship with God.

Three lessons: If ye accept the world thought, ye shall find tribulation; in Truth is peace; you may learn it and find it as I have.

Now, the soul is alone with God and communes only with God. "The hour is come " the consciousness

317

of fulfillment. As the son or individual soul shines with the glory of the Father so is the Father glorified in the work of the Son. There is perfect co-operation between the Universal and the individual. Oneness is the theme of this chapter; first, the soul's recognition of its own unity and co-operation with God, then the recognition of the same truth for all souls.

John 17: 2	The Son's power is of God, hence is unlimited. How does the Son give us eternal life? I John 5:10-12, and "Beloved, now are we the sons of God." Knowing Truth gives us the certainty of Life Eternal. The Son's only work is to glorify good and finish its work.
John 17: 3	Remember Matt. 5:16. God is the glory of the soul, its only light, and that is the eternal light. The soul is of this light eternally; it is born or brought forth in light and
John 17: 4	of light, and this light is its life. John 1:4. The Divine Nature can be manifested or seen only by those who come out of the world thought and obediently accept the
John 17: 5	highest revelation of Truth.
John 17: 6	
	This attitude of mentality can see and does accept the unity of the Father and Son,
John 17:7-8	the Universal and the individual.
John 17: 9	Now, or on this plane where Jesus is speaking, only that which is born of and is like God is acknowledged.

318

These are God's own they have recognized this truth and have turned from the world thought.

A strong affirmation of unity, one Mind, one Life, one Substance, and one Power.	
	John 17: 10
A prayer for a realization of this oneness on the part of the followers of Truth.	
	John 17:11
To be in the world is not to be of the world.	
	John 17: 12, 14, 16
We may be in the midst of worldly opinions, yet not be of those opinions. The son of perdition that is lost can be none other than these very opinions and beliefs, that are	

lost in Truth.	John 17: 13
Our true joy is that which is realized within ourselves and not in another.	John 17: 15
Change of environment will not strengthen us, but to live strong in any environment is the test of strength.	John 17: 17-19
	John 17: 20-21
As it is true of the strongest soul, so is it true of all.	John 17: 22-23
This consciousness is not realized for a few disciples, but for all who believe or will let it be so.	John 17: 24

There is no separation, no partiality, not even for the Son Jesus --- " Thou hast loved them as thou hast loved me."

May all become conscious of having the same relationship to the Father that I have, is Jesus' prayer. Again the affirmation is made of the soul's eternal unity with God "before the foundation of the world."

11

319

John 17: 25	The world thought never recognizes the truth. The soul consciousness knows.
John 17: 26	This consciousness discerns the Divine Nature, and realizes perfect unity of Life and Love. Unity with God is unity with fellow-man, for it is unity with all. It is true companionship.

320

Return to the Top　Return to the Top

PARABLES.

a. "The kingdom of heaven is like unto a certain king which made a marriage for his son.". - p. 321

Parable, a Story; Back of It a Life Lesson

Parables. The reason why Jesus used them. " The kingdom of heaven is like unto a certain king which made a marriage for his son. And sent forth his servants to call them　Matt. 2 2: 2

that were bid- den to the wedding, and they would not come."

This is the first call that the soul hears. It is the voice of Spirit wooing the soul to enter into union with it, but it "would not" heed.

A second invitation is heard; the voice pleads, "Come to me, I will be your supply." . . . " Tell them which are bidden," says the king, "Behold I have prepared my dinner, and all things are now ready. Come unto the marriage." But they made light of it, and went their ways, one to his farm, another to his merchandise. Thought still rejecting the acceptance of Spirit as its supply, thereby slaying, to its own belief, the higher appeal of Truth "and the remnant took his servants and slew them." But when the king heard thereof he was wroth and sent and destroyed those murderers. That condition of thought which rejects the Divine appeal and continues to turn to the external shall finally be destroyed by the Truth.

The third call is received by a very different condition of thought.

"The servants went into the highways and

321

---

gathered together all, as many as they found, both good and bad, and the wedding was furnished with guests."

From this parable we are shown how our thoughts at first refuse to abide in Spirit, because thought is so absorbed in its own chosen ways, seeking satisfaction in the external, killing the gracious messenger of Truth. At last through the lesson of suffering, the call of Spirit is heeded, thought has worked out and proved its own inability to find permanent satisfaction, and is ready to listen to a better way.

Return to the Top Return to the Top

b. The Marriage Supper. . - p. 322

| Luke 14:16-24 | "A certain man made a great supper and bade many, and sent his servant at supper time to say to them that were bidden, come for all things are now ready. And they all with one consent began to make excuse." |

Three excuses were given, showing the slow ness of our thought in accepting the invitation of the Spirit. The first said, "I have bought a piece of ground and I must needs go and see it. I pray thee have me excused." I have turned my attention to the demands of material sense. I am too occupied with these to accept the invitation of Spirit!

And another said, "I have bought five yokes of oxen and I go to prove them. I pray thee have

322

me excused." By five "ties" am I united with the external which I wish to test. May not these five ties or "yokes" refer to the five senses? My senses are all turned to and feeling after satisfaction in the external. I care not to seek from the inner. "Take my yoke upon you." My yoke is union with God. But that which has so many ties to the material, cares not for the one "yoke" which is "easy."

And another said, "I have married a wife and therefore I cannot come." I have become so wedded to, so at one with the flesh, that I cannot. come now into spiritual consciousness.

"Then the master said, go out quickly into the streets and lanes, and bring in hither the poor and maimed and halt and blind. And the servant said, Lord it is done --- and yet there is room.

And the Lord said --- go out into the highways and hedges and compel them to come in that my house may be filled. For I say unto you that none of those men that were bidden shall taste of my supper."

This represents the one standing invitation that is continually offered to each soul.

In the early unfoldment of that soul when it is turning to form as source of satisfaction and is becoming wedded to flesh as something apart from Spirit, thought sees nothing desirable in union with Spirit. That condition of thought, says the

323

---

parable, shall never taste Spirit's supply, But after a while when the external falls, when the stings and suffering of material sense are felt, that same soul, poor and maimed, halt and blind, is glad to come to heed Spirit's call, Come into my fulness.

Return to the Top   Return to the Top

c. The Prodigal.. - p. 324

Luke 15: 11

One son stayed at home, the other wandered neither represented the highest or Christ consciousness. There are thoughts in each one that wander from God and others turn to God, home -- the Source.

The one That stayed at home, the thoughts that stay within, the Christ in us, never wanders. Two sons represent two conditions of thought, belief in two sides, separation. The true father has but one son, the Christ in us.

The younger son here means the one less developed; the Father gave his portion without remonstrance the son asked for little and according to law he received what he asked for, he saw little, so he took little. When we ask for a portion we are not seeing that all is ours, a limited vision. His own act banished him into the "far country," from seeing the fulness. Decree with God and it will be established, because you will see fulness.

Sense of lack (famine) arose in the land of externals. When he found no more

Luke. 15: 15   satisfaction

                        324

in his portion he joined (to one) a citizen of that country, turned still to one who believed in the outer as cause, went into the external and tried new sense beliefs and greater degradation results from experience. Then he remembered his father's house. Man in extremity turns to God. "When he came to himself," the real self, the Christ, he saw the truth of himself and turned at once to his Source, he realized that those who serve in realization have abundance.

"I will arise." We must arise above little things, get a larger vision of our divine Selfhood.
The prodigal then made a determination and carried it out!   Luke 15: 17

It was enough for the father that he had turned, he met him with open arms.   Luke 15: 18

An indolent "stayer at home" has no more claim upon the inheritance than had the   Luke 15: 20
wanderer. And the wanderer returned has proved conscious right. "The first shall be last, and the last first."

The prodigal son was in the presence of God all the time and when he turned away from his own and others' beliefs and opinions, he saw the fulness of all good waiting to receive him. His father had compassion, not human sympathy, pity, but recognition of his son-ship.

Return to the Top  Return to the Top

d. Inportunate widow. Pharisee and Publican.. - p. 326

Importunate widow Pharisee and Publican - Lessons on prayer. We do not pray to please God but to realize the Truth of the Omnipresence of

325

God. Our prayer opens our vision to see all that is ours. Prayer is according to our place of unfoldment and is answered by Law for all is God.

Jesus said, "Ask and ye, shall receive, seek and ye shall find." Seeking is to open our vision to receive, for all is now. Jesus was speaking to children in understanding. He gave another form of prayer to his disciples: "Be ye!" "When ye pray, believe ye have received and ye shall have." The strongest prayer is affirmation; when we question the One Presence by wavering we faint. Recognition is prayer. The parable of the widow is a very low form of prayer.

                        Judge, judged from appearances, he knew no authority but personality, selfishness,
Luke 18: 2   personal sense of self.

Luke 18: 5   When we pray this way we place this judge where we should acknowledge only God.
Luke 18: 9   "Shall not the judge of all the earth do right?" Their view was unjust. Everything from
the God-side is limitless, God never changes.

Parable of Pharisee and Publican shows a false side of prayer, a selfish prayer. They trusted in themselves, the belief in duality is shown here, the two attitudes in each of us.

The Pharisee stands for self-pride, the Publican for humility or humbleness.

The Pharisee said he was not like others (personal pride) he then told of all his good works.

326

We get into heaven not because of our works but because heaven is our home; we are there and we must learn to realize the presence of God everywhere.

The second prayed not the prayer of Truth, for he lowered his thought of himself; he did not lift his eyes to heaven, and he cannot find it until he looks and finds. His motive was not selfish, but it was upon a very low plane of understanding.

Jesus does not say that either prayer is the right attitude, but the Publican was justified rather than the Pharisee.

Luke 10: 30-37

This man turned from Jerusalem to Jericho --fell to lower plane, among thieves. The
result of this .step downward was, that he was stripped of everything. He was walking
by chance, not by principle. The Levite looked and passed by; his was the personal viewpoint. The Samaritan had compassion, that which will help. Oil stands for consecration; wine stands for inspiration. It was in this fashion the Christ in Jesus answered the lawyer's question.

Return to the Top Return to the Top

e. Parable of the Vineyard.. - p. 327

The householder stands for living soul, the I Am. Vineyard stands for realm of individual soul, the field the Master planted. Hired laborers stand for thoughts paid for worthy labor.

Tested their willingness --- time does not count

11a.

327

---

but the attitude does. All are equal but personal thought complained against equality. Problem of personality. The reward of faithfulness is consciousness; seek consciousness for all else is here. Never measure anyone by yourself!

Luke 15:	Parable of the lost sheep and parable of the lost coin teach universal salvation. Saved from ignorance is the only salvation. Saint and sinner are alike to God; He sees the Divine in each. Sinner only sins in ignorant belief the sin is to be destroyed, not the sinner.
Luke 15: 4	

These parables are made to cover a powerful lesson. There is no time limit to salvation, nothing can ever be lost, the Spirit never leaves us.

Shows God's care over all, nothing too small to be included in Universal Salvation. The fold is the consciousness; one sheep shows value of one spiritual thought. Truth has a compelling power. Where is the fold of God? The consciousness within us and everywhere. The Shepherd when he found the one sheep carried it home. I Am is a miniature universe the only salvation is in knowing; rejoice in the Law.

Return to the Top Return to the Top

Help us process more books to support your journey

Click here to give your support to New Thought Library

---

f. The Lost Coin.. - p. 328

PARABLE OF LOST COIN, TEN PIECES OF SILVER.

Luke 15: 8	Money is the world's idol. We have lost to a degree the realization of our true state, and are now finding it through the understanding of
	328

Truth. The process: Brought a light; the lost coin in darkness, ignorance. We need first, Light, which we get from our Source because God is Light. After she had made the light she saw the room and that it needed cleaning, and she swept the house. The coin is our Divinity and it was hid under the dust (personal beliefs) --- the woman searched diligently (no time limit) until she found it!

Return to the Top Return to the Top

Help us process more books to support your journey

Click here to give your support to New Thought Library

---

g. The Rich Man and Lazarus.. - p. 329

The "rich" man and Lazarus represent two stages in unfolding consciousness. Dives, for a time "faring sumptuously every day" while Lazarus begs. The body uppermost, the "spiritual" as it is called in division, receives but "crumbs" of our attention.   Luke 16: 19-31

The picture reverses Lazarus no higher than Abraham's bosom, hears the claim of the brother, to substance, and presumptuously declares him unworthy. Both of these conditions must be lifted up. The individual consciousness in its first awakening to power is dictatorial; as partial understanding always is, and the "body" for a time is the slave of such belief. Only by realization of equality, which is unity, shall such conditions be"done away."

The Scribes and Pharisees are reproved for seeking a sign. "An evil and adulterous generation seek after a sign," a generation that believes in two powers; their belief is adulterated, mixed.   Matt. 12: 38-45

Luke 11: 16, 24,
329   26, 29, 33

There is only one power, God is omnipresent power.

	The truly blessed.
Luke 11:27-28	
Luke 11:37-54	Jesus, sitting at meat with a Pharisee, denounces them, with the Scribes and teachers, as hypocrites, closing his fierce denunciation with, "Woe unto you, lawyers! For ye have taken away the key (the spirit) of knowledge: ye entered not in yourselves, and them that were entering in ye. hindered."
Luke 12:1-59	
Matt. 13:54-58	Jesus' instructions to his followers for all time. The Christ will rule the earth in the hearts of men.
Mark 6: 1-6	
Matt. 10:1, 5-42	Jesus revisits Nazareth, and is again rejected there. "A prophet is not without honour, but in his own country, and among his own kin, and in his own house." And he could there do no mighty work --- and he marvelled because of their unbelief.
Mark 6: 7, 11-13	
Luke 9: 1, 5-6	The twelve disciples are instructed and sent forth. Matt. 11:1. Jesus continues his tour through Galilee. The twelve, including Judas Iscariot, preach repentance, cast out unclean spirits, and heal all manner of sickness and. disease.

Matt.
14: 6-12
> The death of John the Baptist.

Mark 6:
21-29

Mark 6: 30, > The twelve disciples
32-44   return.

Luke
9:10-17

John 6: 1-14   > Thousands are fed on five loaves and two fishes.

Matt. 14: 13-21

330

Jesus walks on the sea many wonderful works of the Spirit are performed at Gennesaret. The people thought the works that Jesus did were contrary to law because they were not according to any law they understood. He did works according to the highest law and it seemed contrary to what they believed to be the natural law. Jesus made

Matt: 14:22-36

Mark 6: 45-56

John 6:15-21

this remarkable statement: "Behold I cast out devils and do cures today and tomorrow, and the third day I shall be perfected." After the third day, that is, he would work no more cures and cast out no more devils. After the resurrection he had a higher work than "casting out," because he was conscious of perfection. He could not have gone on to the ascension if he had seen anything to cure or cast out. This is the very practical reason of his statement, "It is finished." Finished vision, ready for new development or more light. If Divine Science is only teaching us to heal the body,

it is not doing much, but if it is teaching us to get where we do not need healing (and it is doing this great work today), then it is doing the Christ work. The greatest healing is not to need healing!

Jesus' discourse with the multitude in Capernaum in the synagogue of that city and with his disciples. After many of his disciples went back and walked with him no more, Jesus' wistful question was, "Will ye also go away?" This is addressed to the twelve and is answered by Simon

John 6: 22-71

John 7: 1

331

Peter, "Lord, to whom shall we go? Thou hast the words of eternal life. And we believe, and are sure that thou art the Christ, the Son of the Living God."

Return to the Top  Return to the Top

---

PART V. Some events that took place during the twelve months from the beginning of the third Passover.. - p. 332

Some events that took place during the twelve months, from the beginning of the Third Passover.

Matt. 16: 1, 4-6

Mark 8:11-12

The Pharisees and Sadducees again ask for a sign. "O ye hypocrites: ye can discern the face of the sky; but can ye not discern the signs of the times? Take heed, and beware of the leaven of the Pharisees and of the Sadducees." Why doth this generation seek after a sign? "Verily I say unto you, there shall no sign be given unto this generation." No sign was necessary; the Christ in Jesus was doing the perfect work of the Spirit according to law and order.

Matt. 16: 13-20

Peter repeats his confession that Jesus is the Christ. Notice that men believed him to be John the Baptist, Elias, one of the prophets.

Mark 8:27-30

Luke 9:18-21

Matt. 16:21-28

Jesus plainly foretells his sufferings and resurrection. He spake that saying openly, and Peter began to rebuke him. He turned about and looked on his disciples and said to Peter, "Get thee behind me Satan: for thou savourest not the things that be of God, but the things that be of men." He rebuked the personality of Peter, and

Mark 8: 31-38

Luke 9: 22-27

332

---

closed with these wonderful words, " Verily I say unto you, that there be some of them that stand here, which shall not taste of death (what they believed to be death) till they have seen the kingdom of God come with power."

Jesus' transfiguration.

Matt. 17: 1-13

Peter, James and John were with Jesus at the great moments of his life. Like Elisha, who saw Elijah's ascension, they were able to see the transfiguration of this illumined soul. "As Jesus

Mark 9: 2-13

Luke 9: 28-36

prayed the fashion of his countenance was altered, and his raiment was white and glistening.

Behold, there talked with him two men, which were Moses and Elias; they talked of his decease which he should accomplish at Jerusalem." (We have here a very different experience than the one recorded in I Sam. 28:7-20.)

Jesus then admonished his three companions not to speak of this wonderful proof of eternal life they had witnessed, until after he had raised his body from the tomb.

Matt. 17: 9

"This man calleth for Elias."

Matt. 27: 47

The soul that is conscious of eternal life needs no medium through which to enjoy the companionship of loved ones. They will not pass from his vision when he knows that there is no death.

The word itself, in time, will become obsolete; it is without meaning. It is quite natural to think

333

Moses and Elias were deeply interested in the great work that Jesus was doing.

John 18:36-37

John 10: 17-18

" To this end was I born, and for this cause came I into the world, that I should bear witness unto the truth. Every one that is of the truth (conscious of what he had taught them) heareth my voice."

" Therefore doth my Father love me, because I lay down my life, that I might take it again. No man taketh it from me, but I lay it down of myself. I have power to lay it down, and I have power to take it again. This commandment have I received (accepted) of my Father."

A case of healing when the disciples failed.

Matt. 17:14-21

When the child was healed by Jesus, they asked, "Why could not we cast him out?"

Mark 9:14-29

He answered, "Because of your unbelief --- nothing shall be impossible unto you if ye

Luke 9: 37-43   have faith as a grain of mustard seed. Howbeit, this kind goeth not out but by prayer and fasting." Not a distinction as to what is named disease here, but their attention is called to the great necessity of them fasting (from the beliefs and teachings of the world) and praying (the acknowledgment of the power and presence of God as all there is, invisible and visible).

Jesus again foretells his experiences and the resurrection. Continually is he endeavoring

Matt. 17:22-23   to have them understand what was going to take

Mark 9:30-32

334

place trying in every way to prepare them for the great event; the greatest the world has ever witnessed.

Luke 9: 43-45

Who is the greatest in the kingdom of heaven'? Jesus teaches real humility.

Matt. 18:1-35

"Where their worm dieth not, and the fire is not quenched." Matt. 3:12. For there is

Mark 9: 33-50

that in you that will never die; even the hottest flames of Love's refining fire that

Luke 9:46-50

destroys all ignorance, but it, the soul in man, cannot itself be destroyed (quenched). "If thine eye offend thee, pluck it out" --- if your view, though sagacious and intellectually powerful, give it up; better to realize the Universal Presence without your worldly talent than to suffer in the refining flames through which all must pass. God's love is a consuming fire. "If thy foot offend thee " --- if your walk in life interferes with your spiritual unfoldment, abandon that way of living better for you not to feel free from the world's viewpoint, than to hold to certain things that bring you into confusion and unhappiness.

Jesus answers his disciples' questions, "Ye know not what manner of spirit ye are of." Let those who believe in death, bury their beliefs, "but go thou and preach the kingdom of God."

Matt.8:19-22

Luke 9: 57, 62

"No man, having put his hand to the plough, and looking back (giving heed to past opinions and beliefs) is fit for the kingdom of God."

335

Luke 10:1-16   Seventy disciples are instructed and sent out, besides the twelve named in Matt. 10:1-5, Mark 6:7-11, Luke 9:1-5.

John 7:2-53   Jesus goes to Jerusalem; at the feast of tabernacles. He reproves his kinsmen, his brothers, for they did not believe in him. About the midst of the feast Jesus went up into the temple and taught. "My doctrine is not mine, but His that sent me. If any man will do His will, he shall know of the doctrine whether it be of God, or whether I speak of myself. He that speaketh of himself, seeketh his own glory: but he that seeketh His glory that sent him (the Spirit of Truth), the same is true, and no unrighteousness is in him." He then tells them, "Did not Moses give you the law, and yet none of you keepeth the law?"

"Ye both know me, and ye know whence I am: and I am not come of myself, but He that sent me (the Spirit of Truth) is true, whom ye know not." Many of the people believed on him and the Pharisees and chief priests sent officers to take him, but Jesus was not ready to be taken yet, his work was not finished. In the last day, that great day of the feast, Jesus stood up and cried, saying, "If any man thirst let him come unto me and drink." John 4:14. "Whoever drinketh of the water that I shall give him, shall never thirst; but the water that I shall give him shall be in him a well of

336

---

water springing up into everlasting life. Whosoever will may come and drink of the water of life freely." Isa. 55:1. There was a division among the people because of him. The officers declined to take him. "Never man spake like this man," they told the chief priests and Pharisees. Nicodemus, who came to Jesus by night, being one of them, said, "Doth our law judge any man before it hear him, and know what he doeth?" They said unto him, "Art thou also of Galilee? Search and look; for out of Galilee ariseth no prophet." And every man went unto his own house. They were awaiting further developments like the magicians of Egypt, they had discovered the finger of God."

The woman taken in adultery. "Now Moses in the law commanded us that such should be stoned: but what sayest thou?" Jesus stooped down, and with his finger wrote on   John 8: 2, 11-34
the ground. "Man-made law is written upon shifting sands."

Isa. 57: 3-4

Having eyes full of adultery, false teachers speak evil of things they understand not, believe in two standards of living, one for man and another for woman. False teachers,   2 Peter 2: 14
wells without water, clouds that are carried with a tempest. While they promise their followers liberty, the freedom that the consciousness of the truth of God's perfect creation gives to them, they themselves are the servants of corruption (corrupt beliefs,

337

---

a belief in duality, two powers, two laws, when there is only one law and one power -- God's). "Whosoever committeth sin is the servant of sin," man or woman. Every man in this group was convicted of the truth of himself when they went out one by one, beginning at the eldest. They had not a word to say in answer to Jesus' command, "He that is without sin among you, let him first cast a stone at her." But let him who reads not lose sight of the great Teacher's command to the woman in his words, "Neither do I condemn thee: go and sin no more." The woman was given to understand that she had something to do.

Jesus discourses with the Scribes and Pharisees, with those who believed in him and
John 8: 12   with the unbelieving Jews (verse 33). "1 Am (the Christ Spirit in each soul) the light of
the world" (verse 28). "When ye have lifted up the Son of Man, then shall ye know that I Am, and I do nothing of myself; but as my Father (the Source of all that I Am) hath taught me, I speak these things. And He that sent me is with me and ye shall know the truth (the truth of the omnipresence of God, the Father, and your relation to God as his son) and the truth shall make you free. If the consciousness of your sonship shall make you free, ye shall be free indeed. If ye were Abraham's children, ye would do the works of Abraham --- if

338

---

ye were conscious of God, your Father, ye would love me, for I came from God; neither came I of myself, but He sent me." Gal. 4:4. (44th verse.)

Ye have listened to the voice of personal beliefs and opinions, the devil; there is no truth in the claim of anything separate and apart from God.

When he (belief) speaketh a lie, he speaketh of his own, for he is a liar, and the father of it. "Behold, saith the Lord, I fill the heavens and the earth and beside me there is none else." "He that is of God heareth God's words: ye therefore hear not, because ye are not (conscious of the presence of God) of God. (Verse 51.) "If a man keep my saying, he shall never see death." "If I honour myself, my honour is nothing: it is my Father that honoureth me; of whom ye say, that He is your God: yet ye have not known him; but I know him --- and keep his saying. Your father Abraham rejoiced to see my day (my understanding of Truth), and he saw, and was glad. Before Abraham was I Am." John 1:1-2, Ex. 3:14, Coll. 1:15-17, Rev.

1:8.

"Neither hath this man sinned, nor his parents: but that the works of God should be made manifest in him." Truth stood there to heal the blind man. As long as the I Am (the Christ Spirit in each of us) is in the world, the I Am is the light of the world.

John 9: 3

Luke 10: 38-42

Jesus is received into Martha's house.

339

Luke 15:1-32	Jesus defends himself against the Pharisees and Scribes for instructing publicans and sinners.
Luke 16: 1: 31	Pharisees are reproved. "If they hear not Moses and the prophets, neither will they be persuaded, though one rose from the dead."
Luke 17: 11	
Luke 9:51-56	The Samaritans will not receive Jesus. James and John reproved for their zeal against them.
Luke 17:20-37	"Ye know not what manner of spirit ye are of."

The Pharisees ask when the Kingdom of God

should come. Jesus answered, " The kingdom of God cometh not with observation: neither shall they say, Lo here! or, lo there! for, behold, the kingdom of God is within you." Rom. 14:17.

"For the kingdom of God is not meat and drink; but righteousness (right thinking) and peace, and joy in the Holy Ghost." The consciousness of the whole truth, Divine understanding.

Matt. 19: 1-12   Jesus enters Judaea. The Pharisees question him about divorce. The great Teacher refers these men to what we read in Genesis 1:27 and 2:24, the full explanation of

Mark 10:1-12   which is found in lesson G of the Genesis lessons. "Moses, because of the hardness of your hearts, suffered you to put away your wives: but from the beginning it was not so." In the next verse the word "fornication" is used, and as this word means the act of an unmarried person, it can be applied to the married in one sense only, spiritually.

The great soul, conscious of God's perfect creation,

340

---

does not analyze a man-made law; he tells them what was in the beginning, is, and ever shall be: "All cannot receive this saying save to whom it is given --- he that is able to receive, let him receive."

All right-thinking people have believed marriage to be of Divine origin, without understanding its divinity. Just how it was based in the one Source has been ambiguous, lacking definiteness.

All scientific analysis is based in principle all true interpretation of creation and what belongs to it is based in the Creator, where all things have origin and being.

Bodily existence could not be were it not potential in Being before it was brought forth in form.

Universal and formless Substance, Intelligence, Life, Power and Love are omnipresent.

In this Universal Life is ceaseless activity. Of its Substance it forms by its Intelligence it knows and understands how to form; because of its Love it rejoices in forming; through its Power it has all ability to form.

Form is the expression of the Formless, and is wholly dependent upon its Source for its existence.

The Formless is Universal. Form is individual.

341

Body is not the maker or creator of itself, the Creator is self-existent, eternal and always in action; visible form is ever before us, the universe eternally in God.

Men and women must learn the true significance of marriage, its true influence and character, by a knowledge of the Truth of its origin.

Genesis 5:1. Paul's statement that there is neither male nor female in Christ Jesus means that Christ Jesus is God with us expressed in bodily form, in His own image and likeness. Jesus, the expression of God, is the first born of every creature (first to be born into the consciousness of his true Being) and so includes male and female.

Knowing that each one is the same spirit substance, is leaving father and mother and cleaving unto each other. It is Being of my being and existence of my existence, or bone of my bone and flesh of my flesh, what God has joined together.

In Divine Science we are compelled to reverse the edict of human belief that the Creator made a great mistake in providing for reproduction, and we are to let the light of Truth illumine the whole subject; a satisfactory solution of social problems is thus to be evolved.

In Science and Health, Mrs. M. B. G. Eddy says: "Until the spirit of creation is discerned,

342

---

and the union of male and female is seen, as in the vision of the Apocalypse, where a spiritual sense was revealed from heaven, let this union continue".

In Divine Science the spiritual sense of marriage is revealed from heaven, and we understand that there could be no creation, no offspring or expression of life in the creature, without the actual presence of the Creator in man, male and female.

Then shall marriage cease because we know the Truth?

By no means for the Truth reveals that it is eternal in Spirit, and when stripped of all human opinions is wholly spiritual. Jesus' saying, that "In the resurrection they neither marry, nor are given in marriage", means that in Truth the unity of Spirit and body are eternal. It is the presence of the Creator, and His action in man that prompts the true idea of marriage. Jesus did not mean that the symbol should be done away with when he said: "Have ye not read, that He which made them at the beginning, made them male and female. * And they twain shall be one flesh?" He meant that, as they believed in one Spirit just so should they believe in oneness of flesh. Every truth expressed symbolizes the nature of the Expressor of it.

But we should not rely on the symbol as the

343

---

reality constituting marriage, for in the resurrection there are no adverse opinions about it; but instead, "Two are seen as one", and the oneness is what God has joined together.

True marriage is based in unity of Being and upon the recognition of this Truth by both husband and wife. Voluntary co-operation is the result in all that pertains to the welfare of the family; so true union is always based in Spirit.

True independence of both husband and wife is based in their unity and in the fact that they are born free and equal.

In his letter to the Ephesians (chapter five), Paul's idea of submitting ourselves is expressed in the twenty-first verse. After telling them to give thanks always for all things unto God, the Father of our Lord Jesus Christ, "Submitting yourselves one to another in fear (reverence) of God".

Then where he told the wives in the allegory (verse 32, "This is a great mystery: but I speak concerning Christ and the church") to submit themselves unto the husbands as unto the Lord, he meant that we should mentally bow to the Truth, Perfect Being, mentally conform to the Truth that our Being, body and all, is the Christ; and this men and women should ever do to practice Truth.

This is the conception that Mary had of her

344

---

husband when the angel Gabriel appeared to her and the Holy Ghost (the consciousness of the whole Truth) descended upon her.

In this letter Paul conveys no idea of inferiority or superiority between husband and wife. Love and reverence are one; and they two shall be one flesh, side by side in Truth.

What is joined together in God, His Being holds united as one. Now it is clear that marriage is the unity of male and female and that unity is God.

The injunction, "Increase and multiply", is the law of Being. Shall marriage continue? Truth's answer is, it shall never cease to be. "What God hath joined together (what is joined together in God), let no man put asunder". For this reason, in the resurrection they neither marry nor are given in marriage, but are as the angels in heaven. They live according to the Truth of their unity.

Mrs. M. B. G. Eddy says that "Until it is learned that generation rests on no sexual basis, let marriage continue, and let us permit no such disregard of law as may lead to a worse state of society than now exists. Spirit will eventually claim its own, and the voice of physical sense be forever hushed."

It is true that generation does not rest on a

345

---

sexual basis but there is no basis for the sexes but the Creator.

Let all things be based aright, and right results, which are just and loving, will follow. In consciousness of Truth there is no limitation sensed by conforming to the law of the land.

Spirit does claim its own in Truth and there is no physical sense. This true marriage was spoken of by Hosea 2:16-19.

Matt. 19: 16-30 Mark 10:17-31 Luke 18:15-17	Jesus' discourse in consequence of being asked by a rich man how he should attain eternal life. "Why callest thou me good? there is none good but one, that is God but if thou wilt enter into life (consciously) keep the commandments". It is difficult for a rich man to enter into the kingdom of God because he places his love for his possessions before his love for God. But with God, conscious of our unity with God, all things are possible. We unfold into the consciousness of the

presence of God as little children, in perfect, unquestioning faith. The process is from less knowledge to greater knowledge of Truth.

Matt. 20:17, 19,	Jesus, as he is going up to Jerusalem, foretells his crucifixion to the twelve disciples. This is the third occasion upon which Jesus tells in detail the great experience he is

28

Mark 10: 31, 34, 45

Luke 18:31-34

bringing to pass in Jerusalem. He was talking to men who did not understand that marvelous truth of the glory of the son of man to which he was to bear witness

346

before the world. This is clearly shown by the ambitious request of James and John.

Jesus kept the feast of dedication at Jerusalem. As Jesus walked in the temple in Solomon's porch, the Jews said unto him, "If thou be the Christ, tell us plainly". He answered, "I told you and ye believed not; the works that I do in my Father's name (nature) they bear witness of me. I and my Father are one". We believe that only when he could make this declaration, could he prove his consciousness of Omnipresence, Omniscience and Omnipotence; could he with a word transform the world of effect and banish every ill. There was no experimenting, no hesitating, no delay; he fully knew and understood the work he had laid out for himself. There was no need of a betrayer.

John 10:22-39

Jesus went again to Bethabara (John 1:28) after the feast of dedication; and remained there until a fit occasion called him into Judaea.

John 10:40-42

"Lazarus is dead: and I am glad for your sakes that I was not here. To the intent ye may believe;" to Martha, " Thy brother shall rise again" --- and when Martha told him that she knew he would rise again in the resurrection at the last day, Jesus said to her, "I Am (the I Am of each soul) the resurrection and the life: he that believeth in Me (the Christ of each soul), though he were dead (unconscious of the truth), yet shall he

John 11: 1-54

347

live; and whosoever liveth and believeth in me (the Christ in you) shall never die. Believest thou this? Where have ye laid him? Take away the stone."

In reply to Martha, who said that Lazarus had been in the grave four days, Jesus said: "Said I not unto thee, that, if thou would'st believe, thou shoulds't see the glory of God?"

Then they took away the stone and Jesus lifted up his eyes and said, "Father, I thank thee that thou hast heard me. And I knew that thou hear est me always: but because of the people which stand by I said it, that they may believe that thou hast sent me". After this prayer of realization, he could say with intelligence and power, "Lazarus, come forth" --- and, "loose him and let him go"; loosen him from the bondage of belief in death; God is the God of the living. If Jesus wept it was because of their ignorance; his tears could not have been for sympathy because he was prepared to speak the words that would bring to Martha and Mary their brother, whom they believed had gone from them.

John 11: 45,53, 54, 57

Luke 19: 1-28

John 12: 1, 9-11

Meeting of the Sanhedrin; Jesus in Ephraim.

Jesus visited Zaccheus, a chief of the Publicans.

Jesus arrived at Bethany six days before the Passover, and proceeded to Jerusalem, amidst the acclamations of the disciples and of the multitude.

348

The temple, was cleansed. Isa. 56:7, Jer. 7:11. Jesus' discourse with the Priests, Scribes and Elders in the temple.

The Pharisees --- Herodians, the Sadducees, and one of the Pharisees who was a Scribe, questioned Jesus. Jesus questioned the Pharisees.

Ps. 118:22. Stone --- perfected humanity. The Sadducees were confuted. Jesus

Matt. 21:12-13
Luke 19:45-48
Matt. 21: 23-46
Mark 11: 23, 27
Matt. 22: 15, 22
Mark 12: 29, 37

reproved them with Divine eloquence. The Christ is THE Son not A Son. Luke 20:41.

Jesus foretells the destruction of the temple as he takes his final leave of it: and, on the Mount of Olives, teaches four of his disciples what were the signs of the Son of Man's coming with power and great glory --- "Verily I say unto you, this generation shall not pass away till all be fulfilled".

Matt. 24: 1-51
Matt. 25: 1-30
Mark 13: 1-37
Luke 21: 5-36

Return to the Top Return to the Top

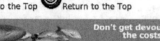

Don't get devoured by the costs

Get an accurate picture of New Thought Education

DivinitySchool.net

PART VI. - The events of three days, from the day of the fourth Passover to the end of the day before the Resurrection. . - p. 349

The events of three days, from the day of the fourth Passover to the end of the day before the Resurrection.

Matt. 26: 17-19

The text shows us very clearly that Jesus had made all necessary preparations for the Passover the Lord's Supper, and from now on, until his work with them was completed, they, especially Judas, followed his directions.

Mark 14: 12-16

Luke 22: 7-13

Judas Iscariot was of South Palestine, the

349

other eleven disciples were of Galilee, this perhaps was the cause of the estrangement between him and and the others. Judaeans had a tendency to look down on Galileans.

The life of Judas previous to his call is hidden from us. (In all the lists in the Gospels his name stands last in the group of four, while in Acts, his place is vacant.)

Matt. 10: 4

Mark 3: 19

Luke 6: 16

Acts 1: 13

Like all of the others, Judas Iscariot received power to cast out demons and heal diseases.

(Matt. 10:1, Mark 6:13, Luke 9:1-6. One writer conjectures that the enthusiast who said, "I will follow thee whithersoever thou goest", (Luke 9:57) was Judas. In connection with their being called to discipleship, we may ask, "Why was such a man as we have been taught to believe Judas Iscariot to be, chosen for a disciple?"

John 2: 24

John 6: 64

Luke 22: 31

Luke 8: 3

John told us that Jesus "knew all men".

What Jesus said of Peter about the same time that he said this about Judas Iscariot. John 6:70.

The company of disciples began to have funds, and Judas carried the bag.

John 13: 29

Verses 27-28. Now no man at the table knew for what intent he spake this unto him, when Jesus said to him, "That thou doest, do quickly". None of the disciples suspected Judas of being dishonest, even at the last moment when this command was given.

350

Return to the Top  Return to the Top

THE LORD'S SUPPER - p. - 351

According to the Syrian custom, they sit on the floor in a circle at this meal. Large dishes containing different kinds of food are set before them; the food is lifted into the mouth with small shreds of thin bread. Even liquid food is sometimes "dipped up" with pieces of bread formed like the bowl of the spoon. Each guest is privileged to reach and dip his bread in the dish.

All dipped in turn in one nearest Jesus. Judas Iscariot was eating in the same fashion as the other disciples he drank wine out of the same cup. The contents of one cup of wine was drunk by the whole company as a seal of their friendship with one another.

"Remembrance" phrases are legion among Syrians; there is never an account of fraternal feasting in Syria complete without the mention of remembrance. A "sop" is a choice bit of food handed only to one of cherished friendship --- a gift of true friendship.

"When he had dipped the sop Jesus gave it to Judas Iscariot and said to him: " That thou doest, do quickly." Now, no man at the table knew for what intent he spake thus unto him, but they thought many things. Judas having received the sop went immediately out, evidently to obey his master's instructions,

12

351

There are only two accounts of the death of Judas Iscariot, and they are very different.

Matt. 27:3-10

The priests bought the potter's field, to bury strangers in. Jeremy? Zechariah 11:12 does not agree with Matthew.

Acts 1:16-20

We are told here that Judas purchased a field with the reward of iniquity. Ps. 69:25. David's prayer was answered. Ps. 109:8. David prayed for help in the midst of his enemies.

Paul's letters were the first Christian documents, yet there is no reference to Judas Iscariot in any of them. His silence upon this subject is worthy of comment.

John 18: 4, 5; 7, 8, 11

Was there need of a betrayer?

Pilate answered, "Thine own nation and chief priests have delivered thee unto me: what hast thou done?"

John 18: 35

Jesus answered Pilate, and in the 38th verse Pilate asked him that greatest of all

John 18:36-37   questions, "What is truth?" The answer is given in the Gospel of Nicodemus, Apocryphal New Testament, 3:11-14. Jesus said, "Truth is from heaven.

Pilate said, "Therefore, truth is not on the earth." Jesus said to Pilate, "Believe that truth is on earth among those, who when they have the power of judgment, are governed by truth, and form right judgment". Reasoning with truth, must give conclusions that are truthful.

352

And when the hour was come, Jesus sat down, and the twelve apostles with him. There was an ambitious contention among the twelve.

Matt. 26: 20
Mark 14: 17

John was present but did not tell of it. " Supper being ended" --- John told of the washing of the feet of the disciples. The more exalted the Christ Spirit the more simple will be our lives. If we have the spirit of the example Jesus gave, in what took place at the supper, and live it, we do not need the letter. The washing of feet would mean the purifying or illumining of the understanding, the cleansing of every false sense.

Luke 22: 14, 24, 30-31

Luke 22:15-18

John 13: 1-20

There are different occasions upon which the supper has figured. John, in Revelation 19:9, heard a voice saying "Blessed are they that are called unto the marriage supper of the Lamb" -- come into unity, which is the true communion.

Matt. 22: 1-13

Luke 14: 15

"Blessed is he that shall eat bread in the kingdom of God." Isa. 55:1-2. "Every one that thirsteth." Prov. 9:16. Wisdom's invitation.

John 6:

How plainly symbolic it is.

"I am that bread of life. I am that bread which came down from heaven. If any man eat of this bread he shall live forever. He that eateth me shall live by me. The bread of God is he that cometh down from heaven and giveth life unto the world." The bread is the substance, the blood is the life. Coming into the consciousness of unity

353

---

with the Presence of God, we partake .of that life and substance which knows no death nor decay.

1 Cor. 11: 23

Told the story of the Last Supper. "For as often as ye eat this bread and drink this cup, ye do shew the Lord's death till he come." This leads us to the second coming --- this coming was expected soon. Jesus had said, "I will come again. I will come to you." They expected it to be fulfilled literally and soon. In Acts 1:11, we read, "This same Jesus which is taken up from you into heaven shall come in like manner as ye have seen him go into heaven." They took this literally, so we can conclude that Paul meant to say, "Take this for a while to remember the Lord until he comes to you again", for lie probably believed that Jesus would come again soon. Paul was the only one who spoke of doing it "until he comes".

"Do this till I come." Symbols of that Presence. Symbols are props to the weakness of personal thought. We feed upon the symbol while we do not realize the Presence of the reality. We cannot use the symbol after the Presence that is with us always, has come to our consciousness. "When that which is perfect is come, that which was in part shall be done away."

Fact: We, the undeveloped, see imperfection where there is only perfection.

Truth: We, the developed, see perfection where we used to see imperfection.

354

God is all the time giving form to his eternal Idea. Form is the eternal manifestation of God.

Shape --- our conceptions take shape, or the undeveloped is always shaping his conceptions, "Law obeyed or disobeyed brings different results."

Shape is imperfect and temporal. The Perfect Mind (God) gives form to its perfect Idea, this perfect form appears to man in the shape of his own conceptions until he sees clearly form is perfect and eternal.

A more spiritual conception is suggested in Hebrews 9:28 --- "Unto them that look for him."

If I look for the Christ, this moment the second coming will be to me. It is a matter of our own mental attitude. Those who are not looking for the coming forth of the Christ spirit from within our own souls will not see it, it is just as far off as we put it.

When we can see the Christ, we shall see ourselves with Him. We shall be conscious of the omnipresence of God as He was and we shall then be able to do the works that he did. Col. 3: 4

1 John 3: 2

That is positive. There is to be no second coming until we are conscious of the Christ in us; when we are like Jesus, he will appear to us. Matt. 26: 31-35

Mark 14: 27-31

Jesus foretells to the apostles the weakness of Peter, and their common danger. He prepares them in a very practical way for what is to follow. Luke 22: 31-38

355 John 13: 36-38

Luke 22:35-38 "When I sent you without purse, and scrip, and shoes, lacked ye anything?" And they said, "Nothing." Then he said unto them, "But now, he that hath a purse, let him take it, and likewise his script: and he that hath no sword, let him sell his garment and buy one. For I say unto you, that this that is written must yet be accomplished in me," then added, "for the things concerning me have an end." When he learned that they had two swords, he said, "It is enough." In the contemplation of the events to follow that night. Jesus knew it would be very likely that his followers would have to cut their way, through the mobs, out of Jerusalem. In verses 40 and 46, Jesus said unto them, "Pray, that ye enter not into temptation." The great Teacher knew that it would be a difficult time for the faithful Galileans who had followed him, because they did not understand the work he was to accomplish. His path was clearly defined, and after the last struggle in the Garden, he never faltered. The Garden of Gethsemane stands for the surrender of the personal, to the universal, or Divine will; that is the struggle we all have to pass through. The Garden is the first step; the last step in this surrender is the crucifixion. It is only personal opinions and beliefs that are crucified. This must all have been true of Jesus. On the cross he cried, as we do over our crosses, "Why bast thou forsaken me?" That cry

356

---

tells the story of every cross; the cross is the way of experience. If we can accept perfection now, we do not need any more crucifixion. The last claim of separation will nail us to the cross, but it leads to resurrection. When Jesus gave up the ghost (the delusion) every sense of isolation was destroyed. Conscious unity with God dawns in resurrection and is brought to full fruition in ascension.

Jesus taken to Calvary, on the cross: who were present during the crucifixion, remaining events of the day. Matt. 27: 32-34-50, 61

Events of the day following the crucifixion.

Mark 15: 21-23, 37, 47

We have been taught to believe that the crucifixion was the aim and purpose for which Jesus lived his life on earth. We have dwelt upon his "death" until we have nearly lost sight of the resurrection. Now what did Jesus say of his own mission? Read John 18:37

and 10:17-18. These statements do not point out anything about death. His crucifixion, outwardly, was the result of the jealousy of the Romans and Jews. Inwardly, it was a necessary step in his unfoldment and to his purpose to give evidence of the truth.

Acts 1:18-19
Luke 23: 26, 33, 46, 56
John 19: 17, 30, 42

Jesus crucified between two thieves. Two thieves represent past and future Jesus represents the present, after we accept the Omnipresence.

Matt. 27
2:62, 66

357

Thief, past, the thing that steals away our present realization; thief, future, steals away our realization of present possession. The malefactor who railed at Jesus represents the past we must drop the past by bringing or calling it into the kingdom. Say to it, " This day (of understanding) shalt thou be with me in Paradise."

The three days in the tomb would mean the process of being perfected in consciousness time had nothing to do with it.

1 Peter 3:18-20

He was preaching to the spirits in prison. This would indicate future salvation, for he was surely preaching the good news of salvation to them.

On the third day came the resurrection, as it always will to the one who has been crucified and has come to the place of illumination. This same resurrection is ours. Rom. 8:11. This spirit dwells in each of us, then if we dwell in it, if we are conscious of it, it will quicken this which we have called the mortal body. We can do the same things Jesus did (John 14:12), but we must know what he knew. His works were the result of the consciousness of what he was; he was one with God, and if he had never varied from that, there would have been no cross. The resurrection was the beginning of the consciousness which was finished in the ascension.

Forty days from the day of Resurrection to the Ascension.

358

Jesus appeared first to Mary Magdalene.

Mark 16: 9-11

Jesus' second appearance.

Matt. 28: 9-10

Jesus, having been seen of Peter, appeared to the two who went to Emmaus. Read carefully the reference given from Luke, the natural conversation between Jesus and the two men "whose eyes were holden." After they had told him about "the things" that had come to pass in Jerusalem, which only a. "stranger would not know," Jesus said unto them, "0 fools, and slow of heart to believe all that the prophets have spoken! Ought not the Christ to have suffered these things, to enter into his glory?" And beginning at Moses and all the prophets, he expounded unto them in all the scriptures the things concerning himself (the Christ).

Mark 16: 12-13

Luke 24: 13-32

1 Cor. 15: 5-6

"Search the scriptures for in them ye think ye have eternal life: and they are they which testify of me (the Christ)." "Ye think," you fancy that eternal life is to be found in the book; no --- but in what the book tells you about, and here I Am as a living example of it. We should read the Scriptures as any other book, for practical instruction.

Isa. 52: 13-15

Jer. 23: 56
Ezek. 34:23-24
Dan. 9:24-27

The Messiah of expectation was not thought of as a man of Divine nature, but as Divinely appointed: All through his career Jesus displayed a reticence and sense of proportion, a spiritual efficiency unequaled among men of all time.

Micah 5: 2

Zech. 12: 10

12a

Zech. 13: 7

359

Mark 16:14-18   Jesus appeared to the apostles in the absence of Thomas.

Luke 24:36-49   Jesus appeared to the apostles, Thomas was present.
John 20: 19-23

John 20:24-29-30

The apostles went into Galilee. Jesus appeared at the sea of Tiberius. " Cast the net on the right side of the ship and ye shall find." The net did not break under the direction of the Christ, the doctrine holds, all that are caught abide forever.

Matt. 28: 16   Other appearances of Jesus after his resurrection.

John 21: 5, 10-12

1 Cor. 15:6-7

Acts 1: 3-8

Mark 16: 19-20

Luke 24: 50-53   Jesus' ascension.

Acts 1: 9-12

The ascension is the last step out of the belief of flesh into the consciousness of spirit; it comes when the soul knows its unity with God. The conscious oneness of the soul with God, the real at-one-ment. Jesus did not make the at-one-ment, he revealed it. At-one-ment is an eternal verity. Ascension is the manifestation of the truth that all is spirit. Between the resurrection and the ascension Jesus could make his body disappear, but it would re-appear he was not fully conscious of the truth of his body all the time. After the ascension he never came back in that form. He could

360

not! The body always follows the soul as did Jesus' body. Every step of the way he went upon the path of his unfoldment, he took his body with him he lifted the body up with the soul, he recognized that there was no soul and body --- but soul body --- one, and that is the Truth he bore witness to before the world, that the body could no more die than the soul. So shall we take up our bodies

by the same spirit that dwelleth in us. We shall see him when we are like him. Jesus disappeared and when they saw the cloud, they thought he was behind it, their opinions. Symbol of cloud is in our vision. There was nothing marvelous in the ascension.

The second coming is when Christ comes to every consciousness as a living reality; and that is our coming to Him. The second coming is God manifest in the race!   John 3: 2

John's conclusion.   John 20:30-31

John 21: 25

361

Return to the Top   Return to the Top

Appearances BEFORE ASCENSION.	Time	Place
1. To Mary Magdalene (Mark 15:9)	Day of His resurrection	Garden
2. To other women from Galilee (Matt. 27:9)	Day of His resurrection	Jerusalem
3. To two Disciples (Luke 24:13)	Day of His resurrection	Emmaus
4. To Peter (Luke 24:34)	Day of His resurrection	Jerusalem
5. To ten Apostles (John 20:19)	Day of His resurrection	Upper Room
6. To eleven Apostles (with Thomas) (John 20:26)	Sunday after His resurrection	Upper Room
7. To seven Apostles and others fishing (John 21:4).	Week following His resurrection	Tiberias
8. To five hundred brethren at once (I Cor. 15)		
9. To James the Less (I Cor. 15)	Unknown	Jerusalem
10. To eleven Apostles and others (John 21:14)	Ascension Day	Bethany
AFTER ASCENSION.		
11. To Stephen at his martyrdom (Acts 7:56)		Jerusalem
12. To Paul at		

| his conversion (I Cor. 15) | | Damascus |
| 13. To the Apostle John (Rev. 1) | Lord's Day | Patmos |

Return to the Top   Return to the Top

ACTS. - p. 363

This book, according to internal and external evidence, was written by Luke, and forms the sequel to his Gospel. It is the history of the foundation and spread of the Christian Church, the former under Peter, Acts 1:12, the latter under Paul, Acts 13:28. It was founded on the day of Pentecost; its first sons were Jews; hence, it appeared only as a Jewish sect in Judaea, and the former part of the book is occupied with its establishment there, with arguments in its favor, and with challenges to disprove the fundamental fact of Jesus' resurrection. Its first development into an organized community, with official staff, provoked the first persecution and martyrdom, which precipitated its extension to Samaria and Syria, and caused a new and more independent center of operations to be planted at Antioch, whence, under Paul (the first converted persecutor) it spread to Asia Minor, Greece, Rome, and various parts of the Gentile world. The motive influence was the direct impulse of the Holy Spirit, not any preconceived plan of an Apostolic body.

In this book all the Articles of the Apostles' creed may be found, chiefly in Peter's speeches. Acts 1:5.

Instantaneous healing. Peter's exhortation.

363

		Acts 2: 4
		Acts 9: 17
		Acts 15: 6, 7, 9
		Acts 3: 2-9

Acts 6:8, 10, 15	Stephen, full of faith and power in the presence of God, spoke words of wisdom.	Acts 3:12-26
Acts 7:	Stephen's address before the high priest.	
Acts 8: 1-3	Introduction of Saul.	
Acts 9: 1, 3-17	Saul's experience.	
	Saul's eyes opened to light; he preached at Damascus.	
Acts 9: 20	Peter taught Cornelius and his company.	
Acts 10:	Peter accused, made his defense.	
Acts 11:	Saul at Antioch; disciples were first called Christians here.	
Acts 11:	Saul, called Paul, could not be justified by the law of Moses. His address at Antioch.	

25-26

	Divine healing.
Acts 13: 9, 16, 38, 39, 44	Deputation to Jerusalem and conference.
	Verse 7 flatly contradicts Paul's own words. Gal. 2:7-14. This conference took place
Acts 14: 8-10	fourteen years after Jesus' resurrection. Gal. 2:1.
	Paul's great sermon. "Whom, therefore, ye ignorantly worship, Him declare I unto
Acts 15:	you."
Acts 17: 22, 28, 29	"In God we live and move and have our being" --
	"We are his offspring."
Acts 18: 3, 4, 5, 7, 14, 20, 24	Paul opposed the Jews he was brought before Gallio, deputy of Achaia, who advised the Jews.
Acts 19: 2, 3, 4, 10, 11, 12	Paul sailed for Syria. Apollos, an Alexandrian Jew, mighty in the Scriptures, knew only the baptism of John. Priscilla and Aquila taught him.
	Paul in Ephesus, found certain disciples who had not heard of the Holy Ghost or of Jesus.

364

They were strengthened by Paul's teaching. For two years he taught in the school of Tyrannus so that all they which dwelt in Asia heard the words of Jesus, both Jews and Greeks.

Vagabond Jews, exorcists mastered, the magicians burned their books.

Acts 19: 19-20

Paul raised the dead.

Acts 20:7-12

At Miletus Paul bade farewell to the elders at Ephesus, his charge to them. Paul staked his life and lost it for the sake of peace with the apostles and to obtain their support.

Acts 20: 17-36

Paul went to Jerusalem, was apprehended, but released when he declared himself a Roman citizen.

Acts 21: 39

Acts 22: 3-21

Paul spoke with pride of his birth and education --- a Jew born in Tarsus of Cilicia, brought up in this city at the feet of Gamaliel, who instructed him according to the perfect manner of the law of the fathers, and was jealous toward God. He

frankly told how he persecuted men and women in good faith until awakened by the voice of Jesus, who led him into the path of obedience to the instruction of the Christ within his own soul. Paul was religious before his conversion, and his contribution to our religion was not shared by his contemporaries, Hebrew and Christian. His conversion did not involve a moral revolution, but of less knowledge to more knowledge. It is said that he was not converted until five years after the resurrection.

365

These years of study proved to be the crucial epoch of his life; it was during this time that he yielded to Jesus' teachings the complete surrender of all treasured by the Hebrews. The power of a trained thinker, a logical reasoner, was Paul's; the province of intelligence is to widen our spiritual vision. We have been born into what he worked for, what he bravely gave his life for, but it would be a grave mistake to try to fit him into the progressive thought of our day. Scriptures must be read with discrimination and processes of knowledge be applied; differences in writers and ages be noted. They should be studied from a basis of knowledge and understanding. With his knowledge of Greek, Paul was able to address his hearers wherever he went without an interpreter. This afforded him an advantage over Peter, who was obliged to use an interpreter in his preaching. Peter spoke Aramaic (Syrian), a widely spread medium of communication in regions of Further Asia.

	"I was free born."
Acts 22: 28	
	"Of the hope and resurrection of the dead I am called in question." The usual dissension
Acts 23: 6,10-11	between the Sadducees and the Pharisees arose and he was taken before the Sanhedrin.
Acts 23: 16	
	Paul's nephew saved his life. To thwart the Jews' plot, he was sent to Caesarea. Paul
Acts 24: 3-7	was brought before Felix.

366

Paul made his defence for his life and doctrine.	
	Acts 24: 10-17
Paul appealed to Caesar.	
	Acts 25: 10-11
When Paul made his defence before King Agrippa, Festus told him that much learning had made him mad.	
	Acts 26: 24
Paul's answer.	Acts 26:25-30
Voyage to Rome; was shipwrecked, but reached the land of Melita safely.	Acts 27:
Very naturally the signs followed the life of integrity to the truth he had accepted and he spent his life in teaching.	Acts 28: 3-7
	Acts 28: 8-9
Healing and teaching go hand in hand. Paul believed in the presence and power of God, in whom we live and move and have being.	Acts 28: 11, 16, 17, 30
The voyage was resumed they arrived in Rome. Paul addressed the leading Jews. He preached in Rome two years.	

Return to the Top   Return to the Top

PAUL'S LETTERS

PAUL'S LETTERS --- - p. 367

Paul's writings were the first Christian documents. He saw first the larger Christ, the Saviour of all humanity, not only of the Jews. Eph. 3:1-11. His letters were written to established churches, as pastor they were the substitutes for his presence; they were not looked upon as Scripture. Each letter was written to meet a particular situation of the church to whom it was addressed, and not for us. Paul used the terms

367

"God" and "Lord" with distinction. God, the Father and Lord, Jesus Christ.

Paul's letters are not in chronological order. First Thessalonians was the earliest and Hebrews the latest.

 Return to the Top Return to the Top

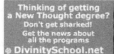 

Romans - p. 368

Paul's letter to the Romans was the only portion of Scripture which approached a system of Christian teaching he was really a great thinker and theologian in the world of his own day. His teachings for the practical life of the Christians were an application of the teachings of Jesus; his realization that love is " the greatest thing in the world," takes us back to Jesus and his teachings. When we compare the words of Jesus in the Synoptic Gospels with the words of Paul in his letters, we see their different view-points. Paul entered into speculations that were not attempted by Jesus, because the theological aspect was pronounced in Paul.

The Apostle's experiences were so intense that it determined in a sense the type of his religion; it made his language lyrical, in a marked contrast to that of Jesus, who so steadfastly taught from a Universal Truth; "I and my Father are one."

Another element in Paul's teaching was his mysticism, his consciousness of the inner light, the Christ spirit within; the reality to him of the invisible. These all constitute phases of religious experience which are growing to be more widespread

368

---

as knowledge becomes consciously universal. The dialectic usage which Paul gave to his religious teachings has an attraction for those who like an intellectual working out of religious principles. As is every great leader, so Paul was a many-sided character. Some of these characteristics have been pushed to an unwise extreme; have been given an emphasis which have distorted Paul's experience and teaching; but for Paul himself, these tendencies were all protected by the intensity of his integrity to the work of giving the message of Jesus to the world.

Our debt to Paul is beyond measure. He was an example of the highest type of religious enfoldment from dogmatic bondage and artificial restraint to complete religious freedom.

Rom. 1: 1, 3, 5, 8

Paul's salutation.

Rom. 1: 8-16

Explanation of his purposes and plans. The Gospel. Rom. 1: 16-18 Because they were vain in their imaginations. Typical pagan life of the time of the early church.	Rom. 1: 19, 24-25

The Jews, in the time of Moses, were in the position of the Gentiles of this day; they had no law; sin, transgression, implies law. Paul is explaining the Divine method of their justification by faith and works as he sees it exemplified in David and Abraham.

Rom. 1: 26-32

Rom. 2: 1, 6, 7, 11, 13, 14, 29

369

Rom. 6: 1, 4, 8, 11, 14, 16	Obedience, the first lesson for the child in spiritual knowledge. "God gives his spirit to them that obey." " To obey is better than sacrifice." I Sam. 15:22.

After hearing Moses' many laws about sacrifice and remembering that the world-thought has been filled with the belief of its necessity, even to the shedding of the blood of an innocent one for the sins of the world, it surprises us to read these words, "1 spake not to your fathers, nor commanded them concerning burnt offerings and sacrifices: but this thing commanded I them, saying, obey my voice and I will be your God and ye shall be my people, and walk ye in all the way that I have commanded you that it may be well unto you." Jer. 7:22-23, Hos. 6:6, Micah 6:6. "Go and learn what that meaneth; I will have mercy and not sacrifice." Matt. 9:13. Read Mark 12:33 and Psalms 50:9-12, and Prov. 21:3, Isa. 1:11-19. "Ye have purified your souls in obeying the truth through the Spirit" --- being born again by the word of God --- a spiritual awakening. I Pet. 1:2223, Heb. 5:8. "Jesus, though a son, yet learned he obedience." II Cor. 10:5. "Bringing into captivity (control) every thought to the obedience of Christ."

Rom. 7:	Inward conflict of Paul. He never followed the personal Jesus he was beyond his time in his development,

370

---

one with the law, but not under it if the Spirit leads.

Rom. 8: 1-2, 11, 16, 17, 18

The law of the Spirit of life frees. "The Spirit itself beareth witness with our spirit, that we are children of God." The glory revealed in us.. 22nd and 23rd --- the redemption of our body. 28th --- to them that love God --- " For whom he did foreknow, he also did predestinate, conformed (in parallel order) to the image of his Son that he might be the first born among many brethren."

We cannot be separated from the love of God. The personal belief and opinion questions God. For Christ (in you) the end of the law for righteousness to every one that believeth.	Rom. 8: 38-39
	Rom. 9: 19-20
All Israel shall be saved. "Present your bodies a living sacrifice (made sacred), holy, acceptable unto God, which is your reasonable service."	Rom. 10:3-4
" There is no power but of God." "Love is the fulfilling of the law."	Rom. 11: 1-26
Passive recognition of our true relation to God is not sufficient there must be the positive attitude that is keenly alive and active. For, "Whatsoever is not of faith is sin."	Rom. 12:1,16-21
	Rom. 13: 1-10
Jesus, who claimed his divinity, who fulfilled the works of the Messiah, was the descendant of David, Jesse's son.	Rom. 14: 22-23
Paul asked their prayers and commended Phoebe unto them; dissensions and apostasy were	Rom. 15: 12
	Isa. 11:10.

371

Rom. 16:

---

warned against and greetings were given to his helpers. Salutations for his companions, concluded with doxology. Written from Corinth A. D. 58 and sent by Phoebe.

Return to the Top   Return to the Top

1 and 2 Corinthians - p. 372

Letters addressed to the church at Corinth, which included those who lived in the adjacent towns of Achaia. Paul was eighteen months at Corinth during his second missionary tour, visiting neighboring cities and establishing centers of work in them. Corinth was the great center of commercial traffic on the overland route from Rome to the east, and also between Upper and Lower Greece. All trade of the Mediterranean flowed through it so that a perpetual fair was held there, to which was added the great annual gatherings of Greeks at the "Isthmian Games," I Cor. 9:24-27. Corinth was proverbial for wealth, luxury and profligacy; population mainly foreign, formed of colonists from Caesar's army and of manumitted slaves, settlers from Asia Minor, returned exiles from the islands, and at this time a large influx of Jews lately expelled from Rome. Acts 18:2.

1 Cor. 1:1-9	Paul's salutation. Reproof of the factions; of personal leaders; exhortation to unity; was sent not to baptize, but
1 Cor. 1:10-17	372

to preach the gospel. Notice the delicacy of touch with which every word is inserted in his writing.   1 Cor. 2: 5

"Your faith should not stand in the wisdom of men, but in the power of God."   1 Cor. 3: 16

"Know ye not that ye are the temple of God, and that the Spirit of God dwelleth in you?"

Return to the Top   Return to the Top

New Meaning of Trinity - p. 373

The great mystery of the Trinity is understood now to mean, not that the mysterious triad exists alone in some desert abyss of space, but that God, the Father, God the Son, and God the Holy Ghost are all there is, invisible and visible.

That man, God and the universe are fused into one is the Christian conception of the Trinity.

To those who accept it in the old way, the idea of the Three in One is an acknowledged puzzle.

Three persons in one God cannot be explained, but those who know God as impersonal Being are solving the mystery.

Christian Science names the Trinity thus:

The Father The Son The Holy Ghost
Life   Truth   Love

More properly speaking, however, these three terms represent the nature of the Trinity: they express the attributes of the Father, Son and Holy Ghost, and are no more to be combined into a Trinity than are other attributes of God, as, Wisdom, Power, Peace,

373

In Trinity, three must complete the One and these three must be inseparable, to which nothing can be added, from which naught can be taken.

Divine Science teaches us the value of the right use of words. The right word in the right place prevents confusion and gives scientific expression to the idea.

We are indebted to Mrs. M. E. Cramer, of California, for the first presentation of this Trinity and the Law of Expression that it reveals. It is the keynote of the teaching of Divine Science and, when understood, is a key to many "mysteries." So freely has this been given to us, that we use it as our own teaching, with additional expressions and changes that have unfolded from its careful study.

The Infinite Being, or Creator, is known to us as the Eternal Mind that is Perfect Intelligence. Creation is known as the bringing forth of that which is forever within the Infinite Mind. Creation is God in self-manifestation. The Infinite is Perfect Order, and its manifestation is governed by law --- its own law.

Trinity explains this Law of Expression. Three is a number that represents perfect law or rule. Trinity reveals the method of Mind's Creation, and the use of this knowledge to us is beyond words.

Mind --- Idea --- Consciousness.

374

This is an analysis of Mind, not a division into parts. Let us consider if these three are indispensable to Perfect Intelligence , if the three make one.

Mind without Idea or Consciousness would be a blank. If you had no idea in your mind, you would lack intelligence and could never accomplish anything. Idea is the starting point of all production; if you paint a picture, write a book, or make an article of any kind, an idea of what you wish to do must first be in your mind, must precede all action.

Consciousness is just as essential. A thousand ideas but no consciousness of them would produce nothing. By consciousness, you know your power and possibility , by the working of consciousness upon your idea you carry out your plan.

Idea and consciousness could not exist without mind to contain them. Then Mind with its idea and consciousness make Perfect Intelligence, and herein is the working power of the Infinite. Since there is but One Mind, One Law, this is true everywhere of Mind and its activity.

The Scientific Trinity presents to us:

Infinite Mind	Infinite Idea	Infinite Consciousness
The Father	The Son	The Holy Ghost

These three are one complete intelligence and perfect power. This is the beginning of all that is "made" , is the bringing forth of the Idea that is in Mind.

375

---

1 Cor. 4: 18-21	"What will ye?"
1 Cor. 6: 20	"Glorify God in your body and in your spirit, which are God's."
1 Cor. 7:	Answer to letter of Corinthian group concerning wedlock, marriage, heathen feasts, public worship, forms that Paul was seeking to establish.
1 Cor. 9: 1-5	Paul accepted and used the Apostolic liberty, but he admonished them not to misuse it. "Am I not an Apostle? Am I not free? Have I not seen Jesus Christ our Lord? Are ye not my work in the Lord?"
1 Cor. 9-5	
1 Cor. 10: 6	"Have we not power to lead about a sister, a wife, as well as other apostles, and the brethren of the Lord, and Cephos?" Paul spoke of the brothers of Jesus without any qualification.
1 Cor. 11:	
1 Cor. 12: 13	Admonition from Israel's history.
1 Cor. 13: 1, 11, 13	Rules for divine worship-26th verse, " Till I come." Till the Christ spirit is consciously born in each one.
1 Cor. 14: 40	Concerning spiritual gifts --- gifts diverse but same Spirit.
1 Cor. 15:1, 5, 8	The greatest of these is Love. God is Love, conscious unity.

Let all things be done decently and in order.

Summary of the Gospel. Jesus appeared last of all to me, as of one born out of due time. 9th and 10th verse --- "I am the least of the Apostles, that am not meet to be called an Apostle, because

376

I persecuted the church of God --- but by the grace of God, I am what I am."

1 Cor. 15: 20

"First fruits of them that slept." The resurrection of Jesus is the pledge of ours. Verse 45 --" The first man Adam was made a living soul, the last Adam, a quickening spirit." Adam, first son of. God; Jesus, first conscious son of God. Verse 55 "0 death, where is thy victory !" Isa. 25:8, Hos. 13:14. Collection for Jerusalem, recommendations. "Watch ye, stand fast in the faith, quit you like men, be strong. Let all your things be done with charity." Paul's salutation: "If any man love not the Lord Jesus Christ, let him be Anathema. Maranatha (0 Lord, come)." Written from Ephesus (A. D. 57) instead of going to them, as he intended.

1 Cor. 16: 13-14

1 Cor. 16:21-22

Return to the Top   Return to the Top

Paul's 2nd Letter to Corinthians - p. 377

This was called for by the effect of the first. In the interval occurred the riot at Ephesus (headed by Demetrius), and Paul's expulsion. Timothy and Titus had both been sent to Corinth, and at. Troas he awaited their return in vain, until he was bowed down with anxiety and fears. The first letter was a careful and logical treatise; the latter was unstudied; it was written in

377

Paul's darkest moments, yet full of exclamations of joy.

	Paul's conscious sincerity. Essentially affirmative.
2 Cor. 1: 8-13	
	Epistle of Christ. Verses 5 and 6 --- " Our sufficiency is of God. Who also hath made us
2 Cor. 1: 19	able ministers of the New Testament, not of the letter, but of the Spirit; for the letter killeth, but the Spirit giveth life."
2 Cor. 3: 2-3	
	Not as Moses.
22 Cor. 3: 13-18	
	Our body.
2 Cor. 5: 1, 5, 7, 16-18	Paul's loving plea. In reverence for God.
2 Cor. 6: 6, 7, 10-12, 16-18	Liberality of Macedonians was an occasion of thanksgiving.
2 Cor. 7: 1	Paul lacked a commanding presence, but was possessed of great courage, tremendous intellectual power, an expression full of grace; at times his face was as the face of an angel.
2 Cor. 8: 9	
2 Cor. 10:17-18	His fear for their loyalty.
2 Cor. 11:	"My grace is sufficient for thee."
2 Cor. 12:2-9	Paul's apostolic credentials.
2 Cor. 12:14-21	"Jesus Christ is in you." Admonitions.
2 Cor. 13: 5	Benediction.
2 Cor. 13:11-14	

Return to the Top Return to the Top

Pauls Letter to Galatians - p. 378

Gal. 1: 1-5

Written at a time when waves of distrust and anger were rolling over his fervent, ardent soul.

Salutation.

378

The Galatians' apostasy surprising.

Gal. 1: 6

"The Gospel preached of me is not after man, not learned in Jerusalem, but by the revelation of Jesus Christ. It pleased God who called me by his grace to reveal his Son in me (not to me) that I might preach him among the heathen. After three years I went up to Jerusalem to see Peter, but other of the Apostles saw I none, save James, the Lord's brother. "Now preached the faith which once he destroyed" --- and they glorified God in me."

Gal. 1: 11-16, 18, 19, 23, 24

Fourteen years. God accepteth no man's person --- gospel of uncircumcision was committed unto me, as the gospel of the circumcision was unto Peter.

Gal. 2: 1

Paul's logical questioning.

Gal. 2: 6-9

Paul spoke with the passion of a heart trembling with anger and anxious love, in order to appeal to his converts who were leaving his teaching; he appealed to them by all that was dearest, by their old affection for him.

Gal. 2: 14

Gal. 3:1, 8, 21, 23, 25, 29

"We, when we were children, were in bondage under the rudiments of the world, but when the fulness of the time was come, God sent forth his Son, made of a woman, made under the law, to redeem them that were under the law --- because ye are sons, God sent forth the spirit of his Son into your hearts --- wherefore thou art no more a servant,

Gal. 4: 1, 3, 4-7

379

---

but a son; and if a son, then an heir of God through Christ."

Gal. 5: 1-14

"Stand fast in the liberty wherewith Christ hath made us free. For all the law is fulfilled in one word Love; thou shalt love thy neighbor as thyself." Lev. 19:18.

Gal. 5: 18

Political imposition is a yoke upon men's necks, but the rule of priests is a fetter to the understanding. If ye be led of the Spirit, ye are not under the law. The Spirit and flesh are contrary one to the other until we understand the truth;

God is spirit that which is born of Spirit is

Spirit, like begets like.

Fruits of the Spirit.

Gal. 5: 22

Individual responsibility. Practical exhortation to use the liberty of the Gospel for the cultivation of true Godliness.

Gal. 6: 4-5

Return to the Top  Return to the Top

Pauls Letter to Ephesians - p. 380

The inscription of this letter is doubtful, thought to have been a circular of which copies were sent to adjacent churches. It is a summary of the revelation of the Spirit as the foundation of the life of understanding expressed in language both fervent and sublime.

Eph. 1: 3-4, 7-8, 10, 17, 18, 23

The blessing of self-revelation, the Christ in us; we are saved by grace --- inward illumination.

Eph. 2: 1-5, 8-10

By grace ye are saved from ignorance. By grace are ye saved through faith, and that not of

380

---

yourselves; it is the gift of God. Not of works lest any man should boast."

"That He might reconcile both unto God in one body by the cross, having slain the enmity thereby --- for through him, by his same nature, we both have access by one Spirit unto the Father." John 14:6.

Eph. 2:11-16

Eph. 3: 8, 14-15

"For this cause I bow my knees unto the Father of our Lord Jesus Christ. Of whom the whole family in heaven and earth is named."

Eph. 4: 4-6

"One body and one Spirit, even as ye are called in one hope of your calling; one Lord, one faith, one baptism, one God and Father of all, who is above all, and through all and in you all."

Eph. 4:13-14

Eph. 4:22-24, 29

"Unto the measure of the stature of the fulness of Christ, that we be no more children, tossed to and fro, and carried about with every wind of doctrine."

Eph. 5: 2, 8-9

Eph. 5: 15-17

"That ye put off concerning the former conversation, the old man (the old way of thinking) and be renewed in the Spirit of your mind; and that ye put on the new man (new attitude), which after God is created in righteousness and true holiness."

"Walk in love, as children of light, for the fruit of the Spirit is in all goodness and righteousness and truth."

"Walk circumspectly --understanding what the will of the Lord is."

381

Eph. 5: 21	"In reverence to God," in the consciousness of the one Presence and Power.
Eph. 5: 22-23	Duties, wives and husbands, children and parents, servants and masters, according to the views of Paul in his time.
Eph. 6: 1-10	"Strong in the Lord, in the power of his might. The armour of God, loins girt about with
Eph. 6: 10, 12, 14-17	Truth, feet shod with gospel of peace, the helmet of salvation, the sword of the Spirit, which is the word of. God." Heb. 4:12.

Make life conform to profession, by conscious unity with God, casting out all feelings leading to discord, by the purity of the Christ within.

Return to the Top ⟳ Return to the Top

Pauls Letter to Phillipians - p. 382

Written during the dark moments of Paul's first imprisonment at Rome. Acts 28.

This letter is full of the gentle and tender charm of his strong character. Deep experience of life is necessary to have full knowledge of Paul's character.

Phil. 1: 1-12	Prayer for their unfoldment in grace.
Phil. 1:18-21	"Christ is proclaimed --- and I will rejoice." "For to me to live is Christ."
Phil. 2: 1-3, 5-11	"Fulfill ye my joy that ye be like-minded --- let this mind be in you which was also in Christ Jesus: who being in the form of God, thought it not robbery to be equal with God; made himself

382

---

of no reputation --- became obedient unto death-- name which is above every name --- the divine nature, that at the name of Jesus, the Christ, every knee should bow --- every tongue should confess --to the glory of God the Father."

"If I am offered upon the sacrifice and service (sacred service) of your faith, I joy and rejoice with you all. For the same cause also do ye joy and rejoice with me."   Phil. 2: 17-18

Warnings against false teachers. "I press toward the mark for the prize of the high   Phil. 3: 3-7, 14 calling of God in Christ Jesus."

"Rejoice in the Lord alway."

Phil. 4: 4

Phil. 4: 7-8, 13

"Peace of God which passeth all understanding shall keep your hearts and minds through Christ Jesus the Christ in you."

"Whatsoever things. are true, honest, just, pure, lovely, of good report, any virtue, any praise, think on these things." "I can do all things through Christ which strengtheneth me."

Parting salutations and benediction.

Return to the Top   Return to the Top

Help us process more books to support your journey

Click here to give your support to New Thought Library

KeysToHeaven.com
The Keys that unlock the door to heaven within

A Spiritual practice is composed of one or more
of the 7 tools of transformation
Another free website to support your journey.

Pauls Letter to Colossians - p. 383

Thanksgiving for their faith, hope and charity, with a prayer for their spiritual unfoldment. The father who path translated us into the kingdom of his dear son. Preeminence of the son, who is the image of the invisible God, the first-born (first-born into the consciousness of sonship) of all creation

13

383

--- the first born from the dead, the first conscious son of God.

Col. 2: 2	That their hearts might be comforted, being knit together in love, and unto all riches of the full assurance of understanding, to the acknowledgment of the mystery of God, the Father, and of Christ, in whom are hid all the treasures of wisdom and knowledge.
Col. 2: 14- 15, 20-22	The ritual law abolished.
Col. 3: 3-4	We shall be like him --- our attitude.
Col. 3: 10-11	"Put on the new man --- Christ in you the glory. Christ all and in all."
Col. 4: 2, 7, 18	Continue steadfastly in prayer. All affairs to be made known by Tychicus. Many loving salutations. With my own hand --- Grace be with you.

Return to the Top   Return to the Top

Pauls First Letter to Thessalonians - p. 384

1 Thess. 1:9	Paul's gratitude for their eager acceptance of the gospel and fidelity in maintaining it, he encouraged them under persecution by his own example.
1 Thess. 2: 4-5, 8, 12, 19	Practical exhortations encouraging cultivation of constructive living. "For what our hope, or joy, or crown of rejoicing? Are not even ye in the presence of our Lord Jesus Christ at his coming? For ye are our glory and joy."
1 Thess. 3: 3, 8, 12	Prayer on their behalf.
1 Thess. 4:9, 11, 13	"As touching brotherly love --- study to be
	384

quiet, to do your own business, to work with your own hands. Not ignorant concerning those who sleep, that ye sorrow, as others which have no hope."

"Ye brethren are not in the darkness --- ye are all the children of light --- let us who are of the day, be sober, putting on the breastplate of faith and love, and for an helmet, the hope of salvation. For God hath appointed us to obtain salvation. Warn them that are unruly, comfort the feeble-minded, support the weak, be patient toward all. See that none render evil for evil rejoice ever more, pray without ceasing, in everything give thanks. Quench not the Spirit, prove all things hold fast to that which is good. Abstain from all appearance of evil. The God of peace sanctify you wholly; your whole spirit and soul and body be preserved blameless. The grace of our Lord Jesus Christ be with you."

1 Thess. 5: 4-5, 8-9, 14-19, 21-23, 28

Return to the Top   Return to the Top

Pauls Second Letter to Thessalonians - p. 385

Affectionate communications and exhortations to perseverance.

"God hath from the beginning chosen you to salvation through consecration of the Spirit and belief of the truth. Therefore, brethren, stand fast, and hold the traditions which ye have been taught whether by word, or our epistle. Comfort

385

---

your hearts, and stablish you in every good word and work."

2 Thess. 3:7-12, 15	Practical way of living, the right use and end of the law of conscious unity. "Be not weary in well doing. Count him not as an enemy but as a brother."

Return to the Top   Return to the Top

Pauls First Letter to Timothy - p. 386

Timothy was the son of a Greek father and a Jewish mother (Eunice) and was converted and circumcised by Paul at Lystra. Acts 16:3.

1 Tim. I:4, 9-15	Recalling charge to Timothy and Paul's claim to his allegiance. Law is not made for a righteous (right thinking) man. Jesus came into the world to save sinners.
1 Tim. 2	Injunctions as to public worship generally, regarding both men and women.
1 Tim. 3	Qualifications of ministers and demeanor of their families. No celibacy.
1 Tim. 4:8-9	Godliness is profitable unto all things.
1 Tim. 4:12, 14, 16	Special advice to Timothy.
1 Tim. 5	Directions respecting communities of widows and elders.
1 Tim. 6: 1-2, 10	To servants. For the love of money (not money itself) is the root of all evil.
1 Tim. 6:11-12, 17, 20	"Thou, 0 man of God, follow after righteousness, godliness, faith, love, patience, meekness. Fight the good fight of faith, lay hold on eternal life. Charge them that are rich in this world that

386

---

they trust the living God who giveth us richly all things to enjoy. 0 Timothy, keep that which is committed to thy trust, avoiding profane babblings, and oppositions of science falsely so called; which some professing have erred concerning the faith. Grace be with thee. Amen."

Return to the Top  Return to the Top

Help us process more books to support your journey

Click here to give your support to New Thought Library

Pauls Second Letter to Timothy - p. 387

Written at Rome in the interval between one trial of Paul before the emperor and that at which he was condemned to death.

"Filled with joy when I call to remembrance the unfeigned faith that is in thee --- for God hath not given us the spirit of fear; but of power and of love, and of a sound mind. God, who hath saved us, and called us with an holy calling, not according to our works, but according to his own purpose and grace, which was given us in Christ Jesus (in the Christ in Jesus, the same Christ in us) before the world began: but is now made manifest by the appearing of our Saviour Jesus Christ, who hath abolished death, and bath brought life and immortality to light, through his teaching and living. Hold fast the form of sound words, in faith and love, which is in Christ Jesus."

2 Tim. 1:
5-7,
9-10,
13

"Be strong in the grace that is in Christ Jesus --- commit thou to faithful men, the things that thou bast heard of me, who shall be able to teach also. If a man strive for masteries, he is not

2 Tim.
2: 2,
5-6, 9,
22-23

13a

387

crowned, except he strive lawfully. The word of God is not bound. Call on the Lord with a pure heart --- avoid foolish and unlearned questions which 'gender strifes.' "

2 Tim. 3:14-17

"Continue thou in the things thou hast learned." All Scripture is given by inspiration of God, and profitable for doctrine, for reproof, for correction, for instruction in righteousness that the man of God be perfect, thoroughly furnished unto all good works." Truth is not man made, but man-discovered. Luke 24:25 --- " Foolish and slow of heart to believe all that the prophets have spoken."

2 Tim. 4:1, 5-7, 9, 18

Personal matters. Death is imminent --"I have fought a good fight, I have finished my course, I have kept the faith. Come shortly unto me.

Return to the Top  Return to the Top

Pauls Letter to Titus - p. 388

Titus, a Greek by birth, addressed by Paul as "Mine own son after the common faith," was the first Christian convert not circumcised. He was taken by Paul to Jerusalem to try the matter, when council decided against its necessity. Gal. 2:3, Acts 15.

Titus 1: 5, 12, 13, 16

Titus was left in charge of church at Crete; a position of peculiar difficulty. Paul advised his deputy upon the course he should take, code of instruction on doctrine, morals and discipline.

388

Sound speech that cannot be condemned. "For the grace of God that bringeth salvation hath appeared to all men."

Titus 2:8-11

"For we ourselves also were sometimes foolish --- not by works of righteousness which we have done, but according to His mercy, by the washing of regeneration; and the renewing of the Holy Ghost which He shed on us abundantly, through Jesus Christ our Saviour." Directions respecting individuals.

Titus 3: 3,5-9

Return to the Top  Return to the Top

Pauls Letter to Philemon - p. 389

Special, to an individual.

Philemon of Colossi, convert of Paul, verse 19. His slave, Onesimus, had run away from him to Rome, having, perhaps, been guilty of misappropriation of his master's goods (verse 18). Falling into Paul's hands, he was converted, reclaimed to his duty, and sent back to his master with this letter of reconciliation. It is remarkable for its generosity and justice. Paul maintains civil rights (even of slavery), confessing that Onesimus, though under the liberty of the gospel, is still the slave of Philemon, and justly liable to punishment for desertion. The damage caused by his absconding, Paul takes upon himself (verse 18). As the returning slave was the bearer also of the letter to the Colossians, it was probably written at the same time, near the close of Paul's first imprisonment at Rome.

389

Return to the Top  Return to the Top

Pauls Letter to the Hebrews - p. 390

The greatest weight of testimony favors the opinion that Paul was author, though probably Luke was writer of this letter. Composed by Paul when he was in strict custody, either at Caesarea or at Rome, just before his death, II Tim. 4:6, when denied writing materials he dictated it to Luke, who then committed it to writing from memory. Some students think we have only a Greek translation of a Hebrew text.

God's word spoken through his Son, the Christ.

Heb. 1: 2-3, 5, 13, 14

The Son whom He hath appointed to be heir of all things --- being the brightness of glory, image of his person. For unto which of the angels said He at any time, "Thou art my Son; this day have I begotten thee, and, I will be to him a Father, and he shall be to me a son. Angels, are they not all ministering spirits, sent forth to minister for them who shall be heirs of salvation?"

Paul's reasoning, why Jesus assumed human nature. Jesus was a divinely natural man.

Heb. 2: 14, 17, 18

The Messiah of expectation was not thought of as a man of divine nature, but as divinely appointed.

Heb. 3:4-6, 19

He that built all things is God, Son, Master of the house; Moses, a servant in it. So we see that they could not enter in because of their unbelief.

Heb. 4: 3, 11, 12, 15, 16

For we which have believed do enter into rest --- although the works were finished from the foundation of the world. Let us labor therefore to

390

enter into that rest, lest any man fall after the same example of unbelief. For the word of God is quick and powerful --- a discerner of the thoughts and intents of the heart. For we have not an high priest which cannot be touched with the feeling of our infirmities but was in all points tempted like as we are, yet without sin. Let us therefore come boldly unto the throne of grace, that we may obtain mercy, and find grace to help in time of need.

"Though he were a Son, yet learned he obedience by the things which he suffered. And being made perfect, he became the author of eternal salvation unto all them that obey him."

Heb. 5: 8-9

Heb. 6: 1-13

Press on into freedom --- Paul exhorted them to be steadfast and diligent.

Heb. 7: 16-28

Story of Melchizidek, who is made, not after the law of carnal commandment, but after the power of an endless life. For the law maketh men high priests which have infirmity but the word of the oath, which was since the law, maketh the Son. who is consecrated forevermore.

For if that first covenant had been faultless, then should no place have been sought for the second. The new covenant of which Jesus was an exponent. That which decayeth and waxeth old is ready to vanish away.

Heb. 8:7-9, 13

Heb. 9: 2.8

The sacrifices of the law inferior to the truth that Jesus taught and lived. "And. unto them

391

---

that look for him shall he appear the second time, without sin unto salvation."

Heb. 10: 11, 16, 17, 19, 20, 22, 24, 31, 34, 36	Which can never take away sins. This is the covenant that I will make with them. By a new and living way. Let us draw near with a true heart, in full assurance of faith. A better and an enduring substance. After ye have done the will of God, ye might receive the promise.
Heb. 11: 1-3 Heb. 12:1, 28	The nature of faith; the faith of the fathers. "Now faith is the substance of things hoped for, the evidence of things not seen." " Through faith we understand that the worlds were framed by the word of God, so that things which are seen were not made of things which do appear."

"Wherefore, seeing we also are compassed about with so great a. cloud of witnesses, let us lay aside every weight, and the sin (ignorance) which doth so easily beset us, and let us run with patience the race that is set before us." Plea for constancy, warning from Esau's case. "Wherefore, we receiving a. kingdom which cannot be moved, let us have grace, whereby we may serve God acceptably with reverence and godly fear, for our God is a consuming fire."

| Heb. 13:1-2, 20 | "Let brotherly love continue." Our privileges and obligations. Now the God of peace, that brought again from the dead our Lord Jesus, that great Shepherd of the sheep --- make you perfect in every good work to do his will, working in you |

392

---

that which is well pleasing in his sight, through Jesus Christ, to whom be glory forever. The Christ in us is the glory --- we have used his personal name instead of his power. Those who were nearest Jesus phrased it "to as many as believed on Him gave He power to become the sons of God." They thought of it as something bestowed rather than revealed.

Return to the Top Return to the Top

LETTER OF JAMES ---

Epistle of James - p. 393

James, brother of our Lord, an Apostle, had the oversight of the church at Jerusalem (Acts 15:13) where he remained until his death.

The epistle is remarkable for its eminently practical nature and for the homeliness and aptness of its illustrations. It was probably written near the close of James' life, and is addressed to the whole "twelve tribes" scattered abroad.

"If any of you lack wisdom, let him ask of God, that giveth to all men liberally. But let him ask in faith, nothing wavering. For let not that man think that he shall receive anything of the Lord.

James 1: 5-7, 17, 18, 22, 25, 27

A double-minded man is unstable in all his ways."

Dual belief. "Every good gift and every perfect gift is from above and --- from the Father of lights, with whom is no variableness, neither shadow of turning. Of his own will begat he us with the word of truth, that we should be a kind of first fruits of his creatures. Be ye doers of the word,

393

---

not hearers only." Pure religion and undefiled is this. (Verse 27.)

James 2: 1, 8, 14, 17, 24

Have not the faith with regard to persons. If ye fulfill the royal law, according to the Scripture, "Thou shalt love thy neighbor as thyself," ye do well. "Faith without works is dead." "Ye see then how that by works a man is justified, and not by faith only."

Of governing the tongue.

James 3:10-11

Wisdom that is from above --- fruit of righteousness is sown in peace of them that make peace.

James 3:17-18

James 4: 8, 13-15

Draw nigh to God. Do not formulate. God is the only intelligence, " The Father knoweth what His children have need of --- it is His good pleasure to give them the kingdom." Luke 12:32.

James 5:1-7, 12, 19

Self-indulgence warned avoid oaths. Pray and praise.

Return to the Top Return to the Top

First Epistle of Peter - p. 394

Simon Peter, son of Jonas, a fisherman at Bethsaida, was one of the foremost apostles, by whom three thousand were converted on the day of Pentecost. Acts 2. And the first Gentile family admitted into Christianity. Acts 10:47-48. He is said to have preached to the Jews scattered throughout Pontus, Galatia, Cappadocia, Asia and Bithynia; that is, the countries of Asia adjacent to the Black Sea, to whom he addressed this epistle from Babylon.

394

Salutation, inheritance incorruptible, that fadeth not away.	1 Peter 1: 1-4, 14
"Be ye holy for I am holy. Vain conversation received by tradition from your fathers. Being born by the word of God which liveth and abideth forever."	1 Peter 1 16, 18, 23
Christ principle in man the chief corner stone.	
"Ye are an elect race, a royal priesthood, a holy nation."	1 Peter 2:7
Be ready to give an answer to every man that asketh you a reason for the hope that is in you. Jesus preached unto the spirits in prison, the gospel preached to those we believed were dead.	1 Peter 2: 10 1 Peter 3:15, 18-19
No longer live to the lusts of men, but to the will of God.	1 Peter 4: 2
So they might live according to the law of God.	1 Peter 4: 6

Return to the Top   Return to the Top

Second Epistle of Peter - p. 395

Peter wrote this epistle when he apprehended his death, II Peter 1:14, and not long after the former one. It is valuable as containing the last words to his converts of one of the original twelve disciples; also for certain personal experiences, such as the mention of the transfiguration by an eye-witness, II Pet. 1:16-18, and the commendation of Paul and his letters. II Pet. 3:15-16.

"Be not idle nor unfruitful. We were eye witnesses of the majesty of Jesus Christ."

2 Peter 1: 8-16

395

---

2 Peter 2:	Peter warns his people against false teachers and bond-servants.
2 Peter 2: 9	The Lord knoweth how to deliver the godly out of temptations.

2 Peter 2:14	Having eyes (vision) full of adultery --- an heart they have exercised with covetous practices.
2 Peter 2: 17-19	These are wells without water, clouds that are carried with a tempest; to whom the mist of darkness is reserved forever. While they promise them liberty, they themselves

are the servants of corruption; corrupt thought, for of whom a man is overcome, of the same is he brought into bondage.

2 Peter 2:21	For it had been better for them not to have known the way of righteousness, than after they have known, to turn from the holy commandment delivered unto them.
2 Peter 3: 2, 9, 10, 13	"That ye may be mindful of the words which were spoken before the holy prophets, and of the commandment of us, the Apostles of the Lord and Saviour." Peter's certainty of Jesus' coming in person. "We, according to his promise, look for a new heaven and a

new earth, wherein dwelleth righteousness. "

Peter's tribute to Paul.

2 Peter 3: 15-16

Return to the Top   Return to the Top

First Epistle of John - p. 396

The tract called the "First Epistle of John" seems rather to partake of the nature of a spiritual

396

discourse, addressed to believers generally, but more particularly to the Gentiles in Asia Minor, probably in the neighborhood of its chief city, Ephesus. Some place the time of its writing before the destruction of Jerusalem, others towards the end of the first century, thinking it bears marks of combating the Gnostic teachings of that time. The theme in this and the following two epistles is Love, conscious unity.

Our fellowship is with the Father and with His Son Jesus Christ.

1 John 1:1-3

"God is light, and in Him is no darkness at all." If we say we have fellowship with Him, and walk in darkness, we are ignorant of the Truth of God and our relation to God. In   1 John 1: 5-8, 10 the light of understanding we do not say we are not ignorant (we have no sin) but we acknowledge our place in unfoldment into the Truth of our Divinity and our Divine possibility of bringing it forth in our living.

"Whoso keepeth His word, in him verily is the word of God perfected he that saith he abideth in Him, ought himself also so to walk, even as he walked. I write no new   1 John 2: commandment unto you, but an old commandment, which ye had from the beginning.   5-8, 14-15, The old commandment is the word which ye have heard from the beginning. Again a   27, 17, 2, new commandment I write unto you, which thing is true in him and in you, because the   25, 29 darkness is past, and the true light now shineth. He that loveth his brother abideth in the light --- your sins are forgiven you for His name's sake. (For the sake of the same divine nature in you). If any man love the world's teachings, the love of the Father is not in him. The world beliefs pass away, but he that doeth the will of God abideth forever. Because ye know the truth, the promise that He hath promised us, eternal life. Everyone that doeth righteousness is born of Him."

1 John 3: 1-2, 4, 8, 9, 11, 13-17, 23-24   "That we should he called the sons of God --- now are we the sons of God and it doth not appear what we shall be but we know that when He shall appear we shall be like Him for we shall see Him as He is." To act in ignorance (sin) is the transgression of .the law --- the penalty of ignorance is suffering. Knowledge of God destroys the works of ignorance. Man does not walk in ignorance when he knows the Truth of his relation to God, his Father. Message heard from the beginning, love one another, conscious unity. The world hates the truth of itself --- hate is the murderer of all joy and freedom that love offers. We lay down the personal belief of our life when we "love our neighbor as we love ourselves." We can only do this when we are conscious of their sonship, their possibilities, as we are of our own. Love one another; we know that the same spirit is in us that was in Jesus.

"Ye are of God --and have overcome the world's teachings, because greater is He (the 16-19, 21   1 John 4: 4, 7, 9-15,

Christ) that is in you, than he (belief and opinion) that is in the world." "Love is of God and everyone that loveth, is born of God, and knoweth God. Jesus manifested the love of God toward us" -- - "Herein is love, not that we loved God, but that He loved us." (Jesus taught us the truth that liberates us from ignorant beliefs. John 15: 22). "No man hath seen God at any time. If we love one another, God dwelleth in us, and His love is perfected in us. (We are conscious of the Presence of God, in whom we live and move and are being.) We know we dwell in Him, and He in us, because He hath given us of His Spirit -- - God is Love and he that dwelleth in love (conscious unity with God means being conscious of our unity with all that has come forth from God, even all that is, invisible and visible.) Herein is our love made perfect, that we may have boldness in the day of judgment; because as he is, so are we in this world! There is no fear in love but perfect love casteth out fear --- his commandment, that he who loveth God love his brother also."

"Whosoever believeth that Jesus is the Christ, is born of God: and every one that loveth Him that begot, loveth him also that is begotten of Him. For this is the love of God, that we keep his commandments: for whatsoever is born of God   1 John 5: 1- 3, 4, 7, 11, 12, 16, 20

overcometh the world" the victory, our faith. For there are three (trinity) that bear record, the Father, the Word (Son), and the Holy Ghost; and these three are one. God hath given to us eternal life. There is a sin unto death (this has been named the "unpardonable sin," the sin against the Holy Ghost). "We know that the Son of God (the Spirit of Truth) is come, and hath given us an understanding, that we may know him that is true. This is the true God, and eternal life."

Return to the Top   Return to the Top

Second Epistle of John - p. 400

This epistle contains only thirteen verses, eight of which are found in substance in the first. It was probably written about the same time, but it is addressed to the elect lady and her children. They are exhorted to persevere in love, faith and godliness and to beware of false teachers.

Return to the Top Return to the Top

Third Epistle of John - p. 400

A short address of a similar kind, to "the well beloved Gaius," of whom nothing certain is known, though he is supposed to be identical with the Gaius named in Rom. 16:23 and I Cor. 1:14. He is commended for his hospitality and piety; warned against the ambition and malice of Diotrephes (one in authority); and his friendly offices are besought for Demetrius.

400

Return to the Top Return to the Top

Epistle of Jude - p. 401

Jude, "brother of James," is supposed to be the Apostle (surnamed Thaddaeus and Lebbaeus) and a brother of Jesus. Matt. 10:3, 13:55, and Luke 6:16.

The epistle is remarkable for the quotation of an otherwise unrecorded saying of Enoch, verse 14, and a tradition of a dispute between Michael, the archangel, and Satan regarding the body of Moses, verse 9. Its date, place and occasion are unknown, but it seems to denounce the same false teachers as those rebuked in II Peter 2, and in very similar language warning them by the examples of the fallen angels of Cain, the impenitent in the times of Noah, of the wicked cities of the plain, of Korah, and Balaam.

Jude, verses 20 and 21 --- "But ye, beloved, building up yourselves on your most holy faith, praying in the Holy Ghost. Keep yourselves in the love of God --- unto eternal life."

Return to the Top  Return to the Top

Revelation - p. 401

The Bible illustrates in various ways the separation of the soul from its Source, the turning of thought from Spirit to a belief in something outside of Spirit. This wandering out of the way of understanding has led thought out of consciousness of Life and Truth.

401

Now we notice that in the first account of this turning away of thought from its Source in Spirit there is no hint of return, but we remember that later, when Israel left Canaan for Egypt, the Divine assurance was, "I will go with thee, and I will bring thee back again." And in the parables of Jesus we learn of the prodigal's departure and his return; of the destruction of the "tares," and the ingathering of the "wheat." It was as if at first man could not grasp such a splendid hope. Eden seemed closed forever when it was said, "And He placed at the east of the garden a flaming sword, which turned every way, to keep the way of the tree of life." Later, man saw better; he now begins to know how he may pass through the sword, and reach the tree of Life; even through the "fire" of purification, for he cannot partake of the tree or source of Life, while he holds a false belief of Life.

Those who have received the idea underlying these lessons are ready for the high call that is sounding through Divine Science today. It is to accept Being wholly. We are this moment within the Source of all, one with its Life, Mind, Substance and Truth, and of its Nature now and forever. This is the Truth that Jesus said we should know --- not a truth outside of ourselves, but the Truth of ourselves. There is but one I Am, but one Principle or Mind that is Being; we are the

402

Individualized parts of the Whole, yet man and woman maintain their identity forever in their individuality, even as numbers which never blend but are governed by one Principle.

Close the eyes to the personal view, look beyond where the Universal rules, and know the truth of I Am. The Universe formless, I see; a vast, boundless space resting in God, the One Intelligence, Its Substance, Spirit, perfect, complete. Here I Am in the Mind that God is and out of this Perfection comes the visible expression I in God and God in me --- whatever God is, I am.

In studying the process of soul unfoldment, increasing individual consciousness, we learn the method of revelation that brings us to consciousness of our Divinity.

In the first "three days" we see the first step in unfoldment. The union of Wisdom and Love brings forth Knowledge. The merest light of Being sets us to action in Love and some degree of knowledge is reached.

The first "three days" is the stage of generation, the conception of bringing forth form from the formless. At its end we have reached the fullness of personal consciousness.

In all generation we shall find the trinity which reveals the Law of Expression, the three essentials of all accomplishment. In it, we find the steps that lead .us forth into form; but later, we

403

retrace these steps as we return to the pure consciousness of Being. In the consciousness of completeness we shall have risen above all conception of cause and effect, beyond the supposition of Creator, creating; we shall see the eternal and changeless manifestation without delay, without the waiting upon time or the development of our ideal.

That we are now able to reason of the process, or conception of Truth, proves that we are nearing the end of conception, which, to understand is to be willing to drop. The end of the process in our thoughts is the end of the world.

We can interpret creation only as we stand above it (in thought). Everything is revealed by the light of understanding. In the process of our unfoldment we say, "In Thy light shall we see light." In consciousness, or identity, we say, "I -am the light."

Every conception, has back of it, a Truth from which it is conceived, the conception of creating form is proof that form is an eternal Truth. Creation is the method by which the soul learns that Truth; it is the opening of the eyes. The first "three days" bring the soul to the knowledge of form; this is only a partial perception of Being, and has but little comprehension of Truth. To full comprehension the soul must rise in understanding.

The first step in this revelation, as also the

404

second, begins with "light" and the decree, "Let there be." The Most High does not decree. To It, Light evermore is. The Truth is that Light fills all; the conception is, that light is just coming into the soul. The latter consciousness says, "I am becoming', the former says, "I am being. I am."

To say, "Let there be," is the beginning of revelation. To the Absolute Truth there is no revelation.

The first step ends with knowledge in which form is seen, lived for, enjoyed. We call this the personal consciousness. Its light is intuition by which we know instinctively, but not logically. We know things but not the reason for things. This condition is described by Paul; "first, that which is natural, then, that which is spiritual."

We decide the length of our "days" and retard or hasten the end of our waiting, according to our reception of the truth we see.

In the highest thought of Creator and Creation, there is, first, Universal Consciousness; second, Individual Consciousness; third, Personal Consciousness.

In the order of development we manifest first, personal; second, individual; and third, universal consciousness. In Truth, the three are one; not so seen though, until we realize the Universal Consciousness,

Personal consciousness presents our first perception of Truth when form is seen without the illumination of understanding. It is a good beginning. It is the coming forth of One into all hence, it is the plane of conception and birth. Death is generated in birth. To conceive of bringing forth in "time" and according to "conditions" is an opinion to which later our consciousness must die. Thus "being born" is the first step to "dying." Generation brings us into the "first death." If we had never been "born" we could never "die"; but because we have been "born, we must die."

As we enter understanding we begin at once to die to what we were, or to what we conceived ourselves to be. "I die daily" --- Paul: Understanding is regeneration, a second birth, re-birth, a higher step. Jesus said, "Ye must be born again." The second step is a "new birth" which we enter through the death of the old. John 8:51. We enter "our world" through birth; we were born into the "earth," we shall leave the "earth" through death (to that belief), and have our new birth in "heaven" (kingdom of harmony), our second step towards full realization of Truth.

In the first, or earth consciousness, body rules.

In the second, or heaven consciousness, soul reigns.

"After death (the first death) the judgment."

As we die to blind faith we have our resurrection in understanding, the day of reason, judgment, discernment. Mark 12:34.

In the first separation we thought God's power apart from man's power, but in the separation made in fuller light, we see this was a misconception and we begin to separate Truth from ignorance, good from evil, wheat from the tares, the sheep from the goats this is our "Judgment" day and it is all right in its place, but there is something beyond.

This is the development of individual consciousness. Man is still looked upon as a dependent existence drawing from a Something above himself, looking beyond himself for the source of possibility and power and Infinite Life, of which he is the "image and likeness." He does not yet comprehend that there is nothing beyond the Infinite Man.

Every "new birth" necessitates a "new death." Having died to our first conceptions we have by the accompanying "birth" risen to higher things, but not to the highest. "Put off the old man, put on the new man." This higher conception and birth must also come to an end. This "end" is called in Revelation the "second death." Rev. 2:11.

Our first step is material conception, our second is spiritual conception, harder to yield than

14

the first. If we will consent to " overcome" (come up over), rise above "heaven and earth," that' are to "pass away," our spiritual as well as our material conceptions will be surrendered to the Eternal. We shall feel no hurt death is but the surrender of personal and individual opinions. The individuality is no less, the individual loves and joys are not taken away, but all are infinitely more, because the sense of self-consciousness of the individuality is taken away, and the individualization is then seen for what it is and delivers thus its full meaning and service in ecstacy to Consciousness. He that loseth his opinions of life shall find the true Life. If we could let go without resistance, there would be no struggle.

The last "three days" in enfoldment complete our second step towards full comprehension. They represent the return of all that has seemed to come forth, to the One. If the coming forth is a conception, so is the return,

Coming forth involves return, but both are less than full consciousness; birth and death are associates, but both disappear in perfect consciousness.

"And God blessed the seventh day because that in it he had rested from all his work."

The highest understanding knows the day in which work is finished to be the most blessed. It is being instead of doing. It is I Am, instead of

408

I am becoming. It is realization instead of hope. It is possession instead of postponement.

"Work days" are good and belong to the place of development. "Rest Day" is better it is the place of realization.

As we study the process let us keep always in thought that it is the showing of the steps by which we pass into full consciousness. We may find therein the mistakes of childhood, we may see the feet that are treading paths slip and sometimes fall, but the trend is always upward.

The study of the process helps us because it explains to us just where we stand in our enfoldment reveals our greater possibilities, and while it makes us willing to "run with patience the race that is set before us," of which no step can be missed, yet it reveals the "end" at hand and, by making us willing to give up misconception for Truth, hastens that end. We find that the "finish of the work" does not depend upon time, but is according to our light.

Revelation (Apocalypse) of John is the only prophetic book of the New Testament and much of it remains unfulfilled.

John, after a vain attempt to put him to death, was banished by Domitian to Patmos but on the Emperor's death he returned, under a general amnesty, to Ephesus, and resumed the supervision of the church there, While in exile he saw and

409

---

recorded these visions, in the introductory chapters of which incidental evidence is furnished that a considerable interval must have elapsed between the foundation of the Asiatic churches and the composition of this book they are reproached for faults and corruptions that do not speedily arise the Nicolaitans had separated themselves into a sect; there had been open persecutions, and Antipos had been killed at Pergamos. Rev. 2:13.

Messages to the seven churches:

1. Ephesus: reproof for forsaking its first love and first works.

2. Smyrna: commendation of works, poverty. endurance of persecution.

3. Pergamos: reproof for false doctrine, immoral conduct, idolatrous pollution.

4. Thyatira: reproof to one party for similar corruptions commendation to the other for their fidelity.

5. Sardis: reproof for spiritual deadness with mere nominal life.

6. Philadelphia: approval of its steadfastness and patience.

7. Laodicea: rebuke for lukewarmness.

These predictions have long been fulfilled; the book is generally regarded as prophetic of the history of the church from the close of the first century to the end of time. By some, the major part is considered to have had its fulfillment in the

410

---

early ages of the church; by others to have been gradually realized by successive religious revivals and persecutions; and again, by others it is regarded as a picture of the historical epochs of the world and the church.

Revelation of Jesus Christ. It is years since

Rev. 1: 2-3

the personal Jesus was here; this is more the Christ Jesus. We acknowledge a perfect man and show how the soul will come into realization. " The time is at hand" means any time one is ready he will hear and keep the Truth. Time refers to man's readiness.

Rev. 1:10, 12-14

The voice behind us shows we have turned away. Seven means perfection, candlesticks mean light bearers. Son of man is not the highest vision. Head means intelligence --- " eyes were as a flame of fire"; if we looked in realization of the Presence of God we should not see evil or ignorance.

Feet of brass stands for the unmixed basis from which we judge, the Omnipresence of God.

Rev. 1: 15

Right hand stands for positive attitude. The seven stars mean individual light. Man in his perfection is saying I Am alive forevermore and I have the key (explanation) of death and hell.

Rev. 1: 16-18

Rev. 1: 20

All churches (candlesticks) of God are Light bearers but the angels (seven stars) bear the message of Truth.

411

| Rev. 12: | This is figurative symbolizing the revelation of Perfect Man. Steps of development to realization. This individual is represented by a woman who is ready to be delivered of a child and a dragon stands ready to devour the child. |

That which brings forth the individual, soul bringing forth the highest it knows.

| Rev. 12: 1 | The woman was centered in light she had ascended to the consciousness of being clothed in Light. The "moon was under her feet," the place we are trying to get beyond, overcoming or putting under her feet. Moon, reflected light, she stands upon all that has not claimed to be the full light of God. Crowned --she was crowned with Intelligence, the full individual intelligence --- this points to no special person; God is |

no respecter of persons it belongs to every individual.

What we bring forth are our children.

Rev. 12: 2

Red dragon stands for earthly conceptions.

Rev. 12: 3

The more we believe in it the larger it grows.

Rev. 12: 4

It stood ready to devour the child earthly conceptions devour our product.

Rev. 12: 5

This teaches us that when we are obedient to the highest Light we know there is a Divine protection to our product. Our product rules our world.

Rev. 12: 6

The woman fled into the wilderness where

412

there is a sense of giving up everything in the silence, alone with God. She was fed forever, numbers beyond all sense limitation.

"War in heaven," struggle within herself. All thoughts of ignorance, all sense of ignorance has deceived us because it has no reality. Rev. 12: 7

Our personal beliefs and opinions are our only accusers. Rev. 12: 9

Spoke the word of Truth, obeyed the call of the soul. Rev. 12: 10

"In the heavens" means Spiritual Conscious ness. The persecution by the dragon Rev. 12: 11
continued, but as long as the woman needed protection she received it.
Rev. 12:12-13

It was much easier for her to go to the wilderness this time than it was in 12:6.
Rev. 12: 14

The flood of personal belief, the human thought that she might be carried away, the
final struggle. Rev. 12:15

Nothing in earth that will harm her. The woman made her unity with all things. Rev. 12: 16

"Blessed are the dead who die in the Lord; they do rest from their labours, and their Rev. 14: 13
works do follow them." Those who die to the personal self and can say, "I no longer live
but Christ (the Infinite Self) liveth" --- have died in the Lord. They "rest from their labours" perceive that
no work that is needed to make me what I Am.

414

"Their works do follow them." Heretofore they have followed the works, looked at results, estimated themselves by their experiences, but having found the real Self, they recognize in man the "Lord of all" and see "works following them" or the results of Being; unfolding as naturally and quietly as the trees unfold in the spring.

415

Return to the Top   Return to the Top

GLOSSARY. . - p. 415

Spiritual Interpretation of Terms

GLOSSARY.

God is Omnipresence, the Presence including all from the least to the greatest, hence, we cannot truthfully claim "error" or "mortality" in anything.

In each and everything is God-Consciousness awaiting man's full realization. From this understanding the following interpretations are offered:

Adam. --- The "earthy" belief, representative not of "error," but of mentality, unfolding as it first awakens into the "living soul." It is a glorious step in that it reveals the rising of man's thought out of his past conception of "dust," and though a small "beginning," it is the surety of complete resurrection. "Being confident of this very thing, that he which hath begun a good work in you will perform it unto the day of Jesus Christ." (Phil. 1:6.)

Angels. --- Divine Thoughts. Every expression of Truth is an angel. Messages of the Creator into its creation --- of the Infinite Mind into its manifestation. Every impulse of Love, every action of right, is an "angel" visitant. Man's pure soul is his angelic Nature.

14a

415

Anointing With Oil. --- " Consecrating to the Truth of Perfect Life, through seeing man's true Being, or Nature."

Allegory. --- Description of one thing under the name of another. A figurative sentence or discourse in which the principle subject is kept from view, and we are left to perceive the intentions of the writer by the resemblance of the picture drawn to the primary subject. A story to illustrate a truth. "Which things are an allegory." (Gal. 4:24.)

Apostles. --- Scholars led by the inner Spirit.

Arm. --- Strength in expressing the Truth.

Atonement. --- Acceptance of unity, an eternal verity. Jesus did not make the at-one-ment, he revealed it.

Baptism. --- Complete immersion in Divine Consciousness. Omniscience. Illumination. Purification of thought.

Being. All that is, both invisible and visible.

Believe. --- To let be that which is and is known.

Blood. --- Circulating Life. Expression of Divine Consciousness.

Blood of Christ. --- Consciousness and Expression of Perfection.

Body. --- Temple of Spirit. "Likeness" of God. Manifestation of perfect Life and Substance.

416

Bones. --- Hidden support of the body; the frame-work of the "temple."

Canaan. --- Land, or Consciousness, of Spirit and freedom as it first appears to thought. Israel enters Canaan and takes possession through the "shedding of blood," because Israel conceives that error has a claim of power even in Canaan, and that Truth must be fought for and bled for. Christians still believe that they can be saved only by the "shedding of blood" taken in its literal sense. True "shedding of blood" is expressing the Consciousness of Love and Life. It is living the perfect Life. "All they that take the sword shall perish by the sword," said Jesus at the last. All conceptions of struggle and fight shall perish.

Christ. --- Son of God. Universal Man. Divine Idea. The Truth that reveals and expresses the Life, Purity, Perfection and Power of the Perfect Mind. The Christ principle is the Son, not a Son.

Conception. --- Man's attempt while limited in belief to form ideas. Conception is yielded when man sees one Idea as the Truth of all things, to be not conceived, but perceived.

Conditions. --- The many and varying experiences of man while he has not found his state of perfect Being.

Day. --- Light. Understanding.

417

Death. --- Destruction of conceptions and misconceptions. "I die daily." This is the death that takes us into "heaven" or spiritual Consciousness, which is perfect harmony.

Disciples. --- Scholars led by personal teacher.

Divine. --- Perfect. Harmonious.

Dust. --- Omnipresent Life and Substance individualized. Conception has called dust materiality, but it is now known that there is no matter, all is spirit.

Ear, Eye, Mouth, Nostrils and Nerves. --- Avenues of soul expression. Faculties that must be enlightened by Consciousness before we can truly judge of the world around us. Unillumined, these senses judge from externals and claim separation of inner and outer. Illumined, they see from within and know the perfect unity.

Earth. --- The manifestation of God now "veiled" by our conceptions. A member of the Universal Body.

Earthquake. --- Upheaval of false conceptions --- a lifting of the "veil." Material belief shaken to its foundations, to be destroyed.

Eden. --- The mentality wherein God is known.

Harmony. The soul's delight in God.

Egypt. --- The external. Rightly r e v e a l e d, Egypt is known as "holy ground." Misunderstood and sought as a Source of supply, it becomes

418

---

to thought a condition of bondage. Our "Israel," spiritual sense, is in captivity in Egypt when our senses are captivated by the outer. "Woe to them that go down to Egypt for help." Only woe can result from looking to the external as a source of help or supply.

Enemies. --- False conceptions. Beliefs in being apart from God. Sin, fear, sickness, and death.

Eternity. --- The Now. All of past, present and future in this present moment.

Eve. --- The first conception of woman. Belief of the separation of male and female.

Faith. --- Understanding; Perfect Trust.

Fasting. --- Abstaining from beliefs of evil. Keeping the "thought from evil and the mouth from speaking guile." A hard lesson, but one that Science helps us to obey by revealing the realities of Life.

Father. --- Source. Life-Giver. All men have one Source.

Fear. --Reverence.

Feet. --Foundation. Understanding.

Fire. --- Purifier. Light, or Consciousness, is the purifying fire.

Flesh. --- In Truth, Spirit-Substance. " The Word was made flesh." Manifestation of Life, but obscured by our ignorance.

419

Flood. --- Omnipresent Consciousness that reveals light filling all. The Baptism of Spirit.

Forgiveness. --- Destruction of False Beliefs.

Gentiles. --- Unreconciled opinions and beliefs.

God. --- Impersonal, Omnipresent, Perfect, One. The Only Being. All Truth, All Substance, Intelligence and Power. Beyond expression, hence, few words can be used in defining God. We can speak of the Nature of this One All as Wisdom, Love, Knowledge, Understanding, Power and Life. This Nature is the Christ of God -- - the Man of God -- that is expressed everywhere at all times.

Grace. --- Inward illumination. Light in the soul.

Glory. --- Out-shining, outward manifestation. When the "Grace of God" fills my thought I shall realize that my body and my world glorify God.

Great Dragon. --- Desire enlarged. (Rev. 12:9, 12.) Its condition just before its destruction. Although desire is a lack of spiritual discernment, it serves a purpose in that it impels man to continue searching until he finds. If, in this search, he mistake the source of satisfaction, he will learn better after many disappointments. Desire still impelling him and even more insistent than ever, man hears, at last the true Voice, accepts the Fullness that is and always has been omnipresent. The work of the " serpent" is ended, when desire

420

---

is satisfied in conscious possession. The Great Dragon represents the great, or intensified, desire of the world for its freedom. Satisfaction is to be found only in Consciousness.

Habitation. --- "Settled dwelling place" of thought. A habit of thought. (Ps. 91:9, 12.)

Hand. --- Power in manifesting Truth.

Hear. --- Understand.

Heaven. --- Spiritual realization, not above us but within us and at hand. The living soul. Eden.

Hell. --- A condition of the mentality that is being purified from self's own thoughts "as by fire."

Holy Ghost. --- Divine Understanding. The Consciousness of the whole Truth.

I Am. The Only One in the Universe. The All-inclusive Mind, Soul, Substance and Life. I am Spirit, living soul and body, for the " three" are the One I am and its manifestation.

Idea. --- Eternal Form in Source of all things invisible and visible. Idea is inherent in Mind and is expressed and made manifest in the unfoldment of Mind called creation. Idea is always perfect and complete. Idea gives form to creation. Thought and body are expressions of Idea.

"Image." --- The mental expression of the Idea that is inherent in Mind. Image is in the mental it is the first imaging of Divine Idea.

421

Israel. --- Spiritual sense not yet conscious of perfection.

Jerusalem. --- Consciousness, veiled by conceptions.

New Jerusalem. --- Consciousness "unveiled."

Jesus. --- The fully awakened mentality. Full consciousness of Being Divine. Complete resurrection of thought. Man knowing himself "lord of all." Ascension of thought above all human conceptions of Mind and body. The realization of unity. The eternal state which man realizes through Consciousness. The destiny of every living soul and body. "As he is, so are we in this world." "The first man is of the earth, earthy; the second man is the Lord from heaven. As we have borne the image of the earthy" (first conception of man --- "Adam"), "we shall also bear the image of the heavenly" (last and true Idea of Man --- "Jesus").

"Kingdom of God." --- State of Harmony "within you." State of all harmonious expression.

"Kingdom of Heaven." Harmonious conditions. The Divine Principle made manifest or visible in earth and body.

"Laying on Hands." Realizing Divine Power.

Life. --- Conscious Being and activity.

422

Life Eternal. --- Life perfect and complete, as well as unending.

"Likeness." --- The living, or visible, manifestation of the eternal Form, or Idea, that is inherent in Infinite Mind. Perfect Body.

Living Soul. --- The image of God. Divine expression. Consciousness expressed as individuality.

Mary. --- " Exalted." The truer Idea of womanhood. Conscious unity of male and female. (Gen. 5:2.)

Man (Universal). --Mind, Conscious Being, Life, Intelligence and Substance. Son of God. The Christ.

Man (Individual). --- Conscious existence. Expression of Divinity. Image and likeness of God. Son of Man. Jesus.

Mind. --- The Supreme Intelligence. The Changeless State containing Perfect Idea and Consciousness. Omnipresence. Its continuous expression is thought and word --- all perfect.

Mountain. Clear vision in realization.

Name. --- Nature. God's changeless Being, expression and manifestation. "In his Name" means because of man's divinity as expression of God.

New Birth. --- Revelation. New realization.

Night. --- Silence. Stillness. The season of

423

---

revelation the in-gathering time for the living soul preparatory to its activity of the day.

Omnipresence. --- One and Only Presence.

Omnipotent. --- One and Only Power.

Omniscience. --- One and Only Knowledge.

Parable. --- A story, back of it a life lesson.

Personality. --- A belief in anything as separate from God. A personal opinion. (John 8:44.)

"Prayer of Faith." --- Acknowledging through understanding. Knowing "you have received." Claiming your birthright of Wisdom, Love, Power and Perfection. Confidence in Truth.

Praying. --- Affirming Truth, and living the Truth affirmed.

Resurrection. --- Evolving mentality. Unfoldment in consciousness of Truth.

Righteousness. --- Right thinking, accompanied by right acting.

Rivers. --- Expression of Spirit's power and possibility. Circulating Consciousness of Life (as blood or the sap of trees). Divine thoughts.

Sacrifice. --- Recognition of Sacredness. Making sacred.

Satan. --- Usurped power. The supposition of a life, substance and intelligence besides Spirit. A lie.

Sea. --- Waters on earth set in bounds. The belief

424

---

that man is separate from God; that earth is divided from heaven, hence, that Spirit's power is limited in the earth or in the body. There shall be "no more sea" when there are no more beliefs of lack and limitation --- no more conception of separation from the Infinite.

See. --- Perceive.

Seed. --- The Divine Nature. The first perception of Truth in man's thought. The beginning of the realization of Wisdom, Love and Life in individual unfoldment.

Serpent. --- Wisdom obscured by ignorance. Desire that urges man to seek until he finds true Wisdom.

Sin. --- False imaginations. (Gen. 6:5.) Ignorance.

Six Days of Creation. --- Unfolding mentality. Fuller and fuller perception of Truth. Self-revelation. Dawning and increasing realization of Light.

Seventh Day. --- Being. End of belief in "becoming." The rest consequent upon knowing the eternal and changeless.

Son of God. --- Perfect Being. Full expression of Truth.

Soul. The eternal Spirit, Substance, Essence. The Truth of all creation. Soul gives Substance to creation.

425

Spirit. --- Substance of the One Mind, omnipresent.

State. --- Eternal and changeless Nature. Absolute Being.

Stone. --- Truth. Foundation. Thought. The Divine Expression. The action of Infinite Mind. The eternal "image" of God. The living soul.

Throne. --- Dominion. Ruling power.

"Tree of Life." --- Man, universal and individual complete as Spirit, living soul and body -- vine, branch and fruit --- all perfect in nature.

"Tree of the Knowledge of Good and Evil." -- The claim that man is a dual nature, subject to two powers, two minds, two wills and two lives. The belief that he must learn by experience and experiment. Man's many opinions that he calls knowledge, blinding his thought for the time to true knowledge.

Truth. --- That which is Changeless and Eternal. The only reality.

The Ungodly. --- Belief of a self apart from God. Any ungodlike conception: A conception destroyed by the flood light of Truth and Love.

Veil. --- False conceptions that blind.

Waters. --- Source of the "Rivers." Spirit's Power and Possibility. Idea of creation that is

426

manifest in all existence. The beginning of expression as it is in Source. (Gen. 1:2.)

White Robes. --- Perfect bodies.

"Windows of Heaven." --- Eyes of the soul.

Word. --- The idea that was in the beginning with God, and was God and --- "was made flesh." (John 1:14.) The Soul of the universe, manifest in the spoken word, deed or body.

Yoke. --- Union. That which unites.

Zion. --- "Raised up." Clear consciousness.

Return to the Top Return to the Top

NUMBERS - p. 427

One. --- Unity. "Constituting a whole."

Three. --- Recognition of unity. Truth expressed by the One, or by the perception of the One as all. " The third day I shall be perfected." One represents Spirit, all inclusive; two represents activity of the one, expansion in thought; three represents the body known in unity with Spirit. The "third day" is therefore the resurrected thought of the body.

Six. --- Double three. Three begins the resurrection of thought; six finishes the seeming unfoldment.

Seven. --- Rest realized after the supposition of a long process ended. Ignorance forgiven by knowledge gained. The thought of man enters into its eternal resting place. Its going forth and

4

427

coming back have but brought it to the realization of Source in Spirit. "In returning and rest shall ye be saved." (Isa. 30:15.) The result of all work (six days) is to bring mentality to the realization of the eternal in which there never was a "process."

Twelve. --- Perfect foundation of any system. As twelve sons of Jacob twelve disciples twelve foundations to the "Holy City."

Forty. --- A process begun and completed. The Israelites wandered forty years. Moses was forty days on the Mount. The "flood" lasted forty days. In "forty days" the "resurrection" was finished in the "ascension." Combinations of these numbers intensify their meanings, as, "seventy times seven," entire forgiveness.

NOTE-Words are often used in two ways in the Bible; used to express conception and Truth.

Example: Psalm 24:1-"The earth is the Lord's and the fullness thereof-." "Earth" is Truth here.

Genesis 6:12 and Genesis 7:21-"God looked upon the earth and it was corrupt and all flesh died that moved upon the earth. "Earth" is conception here.

Job 19:25, 26, 27-Job saw the Truth when he said: "Yet in my flesh will I see God." Flesh is incorruptible from the basis, the Omnipresence of God.

Psalm 65:2-"O thou that hearest prayer, unto thee shall all flesh come." All flesh shall be known as divine.

Galatians 5:17-"The flesh lusteth against the Spirit, and the Spirit against the flesh." Conception warreth against Truth.

428

Return to the Top Return to the Top

INDEX.. - p. 429

429

430

431

VEIL - 163
VOICE - 210,245,292, 411

WATERS - 243
WEDDED - 323
WHERE AND WHAT IS THE BEGINNING- 16
WHY ALL IS GOOD - 58
"WHY HAST THOU FORSAKEN ME" - 356
WORD - - 100

YOKE 323

Return to the Top  Return to the Top

Did you enjoy this Book?

Go to our donations page and send a donation today to support this work!

home | about us | new additions | great links | audio Books | Terms Of Service | contact Serving NewThought |